Compendium of Grape Diseases

Edited by
Roger C. Pearson and Austin C. Goheen

APS PRESS
The American Phytopathological Society

Financial Sponsors

American Cyanamid Company
American Society for Enology and Viticulture
BASF Wyandotte Corporation
CIBA-GEIGY Corporation
E. I. du Pont de Nemours & Company
Eli Lilly and Company
Imperial Chemical Industries
Jeffersonian Wine Grape Growers Society, The
Mobay Chemical Corporation
Monsanto Agricultural Company
Montedison USA, Inc.
New York Wine and Grape Foundation
NOR-AM Chemical Company
Pennwalt Corporation
Rhône-Poulenc Inc.
Rohm and Haas Company
Stauffer Chemical Company
Uniroyal Chemical Company

Front cover: Chardonnay cluster; photo by T. J. Zabadal, Cornell University Cooperative Extension, Penn Yan, NY
Back cover (clockwise from upper left): Clusters of Isabella (photo by T. J. Zabadal), White Riesling (photo by T. J. Zabadal), Cabarnet Franc (photo by R. C. Pearson, New York State Agricultural Experiment Station, Cornell University, Geneva), and Ruby Seedless (photo by F. Jensen, University of California, Kearney Agricultural Center, Parlier)

Library of Congress Catalog Card Number: 88-070733
International Standard Book Number: 0-89054-088-8

Second printing, 1990
Third printing, 1994
Fourth printing, 1998

Printed in the United States of America on acid-free paper

The American Phytopathological Society
3340 Pilot Knob Road
St. Paul, Minnesota 55121-2097, USA

Preface

This compendium of diseases and disorders of grape was compiled as a reference for growers, viticulturists, crop advisers, and extension specialists as well as practicing plant pathologists. It was written for those with limited training in viticulture and pathology but has sufficient detail to be useful to students and professionals alike.

Each section was written by an active researcher or recognized expert on the subject. Included among the more than 40 authors are some who have not previously published in the English language, thereby making their knowledge available for the first time to English readers.

The structure and growth stages of the healthy grapevine are described in the Introduction in addition to a discussion of grapevine diversity that includes the important species of *Vitis* that have been used in breeding programs and as rootstocks. In subsequent sections, diseases are arranged according to causal agents. Diseases caused by biotic agents such as fungi, bacteria, viruses, and nematodes are presented in Part I. Diseases caused by fungi are further subdivided according to the major part of the plant affected. Part II covers some forms of mite and insect injury that could be confused with diseases. Disorders caused by abiotic factors, such as nutrient deficiencies, environmental stresses, and chemical toxicities, are the subject of Part III. Part IV discusses how cultural practices influence disease. Efforts to prevent disease spread in this clonally propagated crop through the selection of disease-free planting material are described in Part V, including clonal selection, clean stock registration and certification, and quarantine. A list of equivalent disease names in French, German, Italian, and Spanish is presented in the Appendix. The Glossary of both viticultural and plant-pathological terms should aid all readers.

Because of the international scope of this compendium, principles of disease control were purposely stressed rather than specific chemical control measures. Nevertheless, in sections where it seemed desirable, chemical control options are discussed in some detail.

We are indebted to many individuals at the New York State Agricultural Experiment Station, Geneva: Rose Sticht for drawings and photographic services; Joe Ogrodnick for photographic expertise and services; Bernadine Aldwinckle for photographic services; Carol Gnau, Carol Gonsalves, and Muriel Stobie for typing the manuscript; and Charlotte Pratt for compiling many of the viticultural terms for the glossary. We are sincerely grateful to the following individuals for translating the disease names from English into their native languages: Bernadette Dubos and Adrien Bolay (French), Wilhelm Gärtel (German), Giovanni Martelli (Italian), and Daniel Téliz (Spanish). We are indebted to Richard P. Korf, Plant Pathology Herbarium, Cornell University, for checking fungal names and to Joe Fulton, David Gadoury, Fred Jensen, and Beth Teviotdale, who reviewed the entire manuscript.

We gratefully acknowledge the following individuals, who authored sections of the compendium and in some cases also reviewed parts of the manuscript:

M. Bisiach, Universita degli Studi della Tuscia, Viterbo, Italy
J. Bulit, Institut National de la Recherche Agronomique, Centre de Recherches de Bordeaux, Pont-de-la-Maye, France
T. J. Burr, New York State Agricultural Experiment Station, Cornell University, Geneva
M. V. Carter, The University of Adelaide, Waite Agricultural Research Institute, Glen Osmond, South Australia
A. Caudwell, Institut National de la Recherche Agronomique, Dijon, France
L. P. Christensen, University of California, Kearney Agricultural Center, Parlier
M. Clerjeau, Institut National de la Recherche Agronomique, Centre de Recherches de Bordeaux, Pont-de-la-Maye, France
B. Dubos, Institut National de la Recherche Agronomique, Centre de Recherches de Bordeaux, Pont-de-la-Maye, France
D. L. Flaherty, University of California Cooperative Extension, Visalia
J. A. Foster, U.S. Department of Agriculture, Animal and Plant Health Inspection Service, Plant Protection and Quarantine, U.S. Plant Quarantine Facility, Glenn Dale, Maryland
P. Galet, Ecole Nationale Supérieure Agronomique de Montpellier, France
W. Gärtel, Biologische Bundesanstalt für Land- und Forstwirtschaft, Institut für Pflanzenschutz im Weinbau, Bernkastel-Kues, Federal Republic of Germany
A. C. Goheen, University of California, Davis
D. Gonsalves, New York State Agricultural Experiment Station, Cornell University, Geneva
J. Granett, University of California, Davis
T. Herrera, Centro de Investigaciones Agricolas del Norte, Campo Agricola Experimental de la Laguna, Torreon, Coah., Mexico
W. B. Hewitt, University of California, Davis
D. L. Hopkins, University of Florida, Leesburg
G. L. Jubb, Jr., University of Maryland, Keedysville
A. N. Kasimatis, University of California, Davis
R. Lafon, Institut National de la Recherche Agronomique, Centre de Recherches de Bordeaux, Pont-de-la-Maye, France
P. Larignon, Institut National de la Recherche Agronomique, Centre de Recherches de Bordeaux, Pont-de-la-Maye, France
J. Lehoczky, Plant Protection Institute, Hungarian Academy of Sciences, Budapest, Hungary
L. S. Leu, Taiwan Agricultural Chemicals and Toxic Substances Research Institute, Taichung Hsien Taiwan, Republic of China
S. D. Lyda, Texas A&M University, College Station
G. P. Martelli, Universita Degli Studi di Bari, Bari, Italy
J. R. McGrew, Hanover, Pennsylvania
R. D. Milholland, North Carolina State University, Raleigh
S. M. Mircetich, U.S. Department of Agriculture, University of California, Davis
I. I. Mirică, Staţiunea de Cercetare şi Producţie Viti-vinicolă, Argeş, Rumania
L. T. Morton, Broad Run, Virginia
S. Nelson-Kluk, Foundation Seed and Plant Materials Service, University of California, Davis
C. G. Panagopoulos, Athens College of Agricultural Sciences,

Athens, Greece
R. C. Pearson, New York State Agricultural Experiment Station, Cornell University, Geneva
F. G. Pollack, Naples, Florida
R. M. Pool, New York State Agricultural Experiment Station, Cornell University, Geneva
C. Pratt, New York State Agricultural Experiment Station, Cornell University, Geneva
R. D. Raabe, University of California, Berkeley
D. C. Ramsdell, Michigan State University, East Lansing
D. J. Raski, University of California, Davis
V. Savino, Universita Degli Studi di Bari, Bari, Italy
W. C. Schnathorst, University of California, Davis
H. Schöffling, Landes- Lehr- und Versuchsanstalt für Landwirtschaft, Weinbau und Gartenbau, Trier, Federal Republic of Germany
H. Schüepp, Eidgenössische Forschungsanstalt für Obst-, Wein- und Gartenbau, Wädenswil, Switzerland
G. Stellmach, Biologische Bundesanstalt für Land- und Forstwirtschaft, Institut für Pflanzenschutz im Weinbau, Bernkastel-Kues, Federal Republic of Germany
L. H. Weinstein, Boyce Thompson Institute for Plant Research at Cornell University, Ithaca, New York
W. F. Wilcox, New York State Agricultural Experiment Station, Cornell University, Geneva
R. N. Williams, Ohio State University, Ohio Agricultural Research and Development Center, Wooster
L. T. Wilson, University of California, Davis

We also wish to thank the following individuals, who reviewed sections of the manuscript:

H. O. Amberg, Grafted Grapevine Nursery, Clifton Springs, New York
R. O. Barrett, New York State Department of Agriculture and Markets, Albany
P. Bertrand, University of Georgia, Tifton
A. Bolay, Station Fédérale de Recherches Agronomiques de Changins, Nyon, Switzerland
A. J. Braun, New York State Agricultural Experiment Station, Cornell University, Geneva
B. Comeaux, East Texas State University, Commerce
T. J. Dennehy, New York State Agricultural Experiment Station, Cornell University, Geneva
M. A. Ellis, Ohio State University, Ohio Agricultural Research and Development Center, Wooster
J. P. Fulton, University of Arkansas, Fayetteville
D. M. Gadoury, New York State Agricultural Experiment Station, Cornell University, Geneva
M. C. Goffinet, New York State Agricultural Experiment Station, Cornell University, Geneva
W. D. Gubler, University of California, Davis
M. B. Harrison, Cornell University, Ithaca, New York
R. B. Hine, University of Arizona, Tucson
Q. L. Holdeman, Department of Food and Agriculture, Sacramento, California
F. Jensen, University of California, Kearney Agricultural Center, Parlier
J. J. Kissler, University of California Cooperative Extension, Stockton
J. J. Marois, University of California, Davis
M. V. McKenry, University of California, Kearney Agricultural Center, Parlier
K. Mendgen, Universität Konstanz, Konstanz, Federal Republic of Germany
L. W. Moore, Oregon State University, Corvallis
R. C. Musselman, University of California, Statewide Air Pollution Research Center, Riverside
J. Northover, Agriculture Canada, Vineland Station, Ontario
R. L. Perry, Michigan State University, East Lansing
R. Pezet, Station Fédérale de Recherches Agronomiques de Changins, Nyon, Switzerland
J. W. Pscheidt, New York State Agricultural Experiment Station, Cornell University, Geneva
A. H. Purcell, University of California, Berkeley
D. J. Rogers, De Funiak Springs, Florida
I. C. Rumbos, Plant Protection Institute, Volos, Greece
M. N. Schroth, University of California, Berkeley
N. J. Shaulis, New York State Agricultural Experiment Station, Cornell University, Geneva
E. F. Taschenberg, New York State Agricultural Experiment Station, Cornell University, Fredonia
D. Têliz, Colegio de Posgraduados, Chapingo, Mexico
B. Teviotdale, University of California, Kearney Agricultural Center, Parlier
J. A. Wolpert, University of California, Davis
E. I. Zehr, Clemson University, Clemson, South Carolina

Additional photographs were supplied by the following individuals:

J. Clark, University of California, Davis
J. A. Cox, North East, Pennsylvania
M. E. Daykin, North Carolina State University, Raleigh
R. H. Gonzalez, Universidad de Chile, Santiago, Chile
R. Guggenheim, University of Basel, Switzerland
J. Harvey, U.S. Department of Agriculture, Fresno, California
A. Kudo, Fruit Experiment Station, Kuriyagawa, Morioka, Japan
A. J. Latham, Auburn University, Auburn, Alabama
L. V. Madden, Ohio State University, Ohio Agricultural Research and Development Center, Wooster
A. H. McCain, University of California, Berkeley
J. Schlesselman, Reedley, California
B. C. Sutton, Commonwealth Mycological Institute, Kew, Surrey, England
J. K. Uyemoto, U.S. Department of Agriculture, University of California, Davis
R. Wind, Bundesforschungsanstalt für Rebenzüchtung Geilweilerhof, Siebeldingen, Federal Republic of Germany
T. J. Zabadal, Cornell University Cooperative Extension, Penn Yan, New York

Contents

Introduction

The grape is the most widely planted fruit crop in the world, covering an area of approximately 10 million hectares. It grows from temperate to tropical regions, but most vineyards are planted in areas with temperate climates. The most concentrated cultures are in Europe. In 1983 the United States ranked seventh in grape acreage worldwide, following Spain, the Soviet Union, Italy, France, Turkey, and Portugal, in that order. Grapes are also grown widely in countries of the Southern Hemisphere.

The grape is a crop plant of many uses. Its fruit is fermented to wines and brandy. It is also eaten fresh and, through the use of cold storage and production in the Southern Hemisphere, is available throughout the year. The fruit is also dried into raisins. Nonfermented juice, frozen juice concentrate, and preserves are common uses for the fruit in North America and other markets. In 1982, the commodity value of grapes for fresh and processed uses in the United States was $1.34 billion, second only to potatoes among all fruits and vegetables.

Diseases in grapes, as in other crop plants, can result in substantial losses in production. In most cases, disease is the result of an interaction between a susceptible host and a living pathogenic organism. Causal agents that are living organisms, such as fungi, bacteria, viruses, and nematodes (Fig. 1), are called biotic, infectious, or parasitic pathogens. Disorders appear similar to diseases but are caused by nonliving entities, such as a nutrient imbalance, environmental stress, or chemical toxicity; these disorders are often referred to as abiotic, noninfectious, or nonparasitic diseases. The disease complex in each viticultural region depends on the array of particular causal agents that are present, the susceptibility of the particular grape cultivars to the causal agents, and the climate of the region.

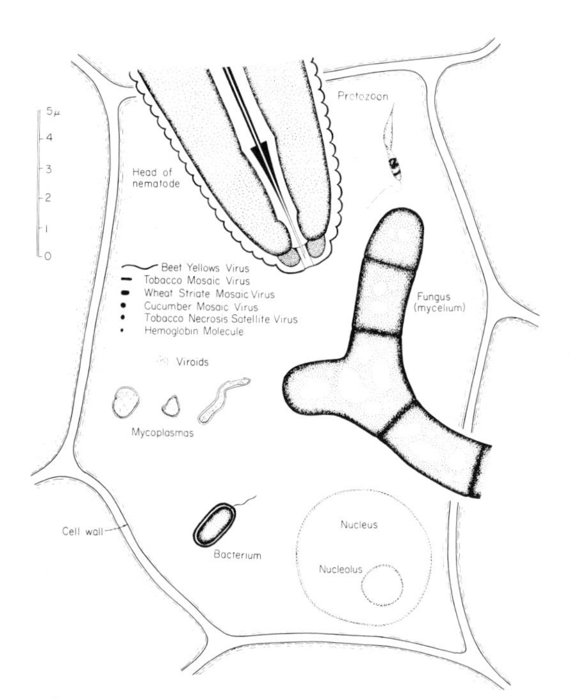

Fig. 1. Schematic diagram of relative sizes of various plant pathogens compared with a plant cell. (Reprinted, by permission, from Agrios, 1978)

Pathogens of Grape

Fungi

Fungi are small, frequently microscopic, multicellular or filamentous organisms that lack chlorophyll. Lacking chlorophyll, they must derive their energy either from dead animal or plant materials, as saprophytes, or from living animals or plants, as parasites or pathogens. Fungi may infect plants through direct penetration of tissue or through an injury or a natural opening, such as stomata, hydathodes, and lenticels.

Bacteria

Bacteria are microscopic, rod-shaped or spherical, prokaryotic organisms. They survive from one season to the next in or on soil, plant debris, seeds, insects, or infected plants, and they are spread by splashing rain or running water, wind, insects and other animals, or humans (by transport of soil, infected plant material, or pruning tools). They enter plants through wounds or natural openings (stomata, hydathodes, lenticels). Some bacteria are confined to certain tissues in the plant, such as xylem vessels, and are transmitted by leafhoppers or by grafting. Bacteria are usually identified by physiological, biochemical, and serological tests.

Mycoplasmas

Mycoplasmas are small, sometimes submicroscopic organisms similar to bacteria except that they have a single, three-layered cell membrane and lack a true cell wall. They vary in shape and are usually nonmotile. They are generally associated with diseases of the yellows type that induce dysfunction of the phloem. They are transmitted by leafhopper vectors and by grafting.

Viruses

Plant viruses are submicroscopic particles usually consisting of ribonucleic acid (RNA) wrapped in a protein coat. Virus particles are polyhedral or rod-shaped (from short and rigid to long and flexuous). Viruses do not reproduce themselves; instead, they induce the host to produce more virus particles. Viruses are transmitted by insect or nematode vectors, by mechanical contact through wounds, and by humans through propagation of infected cuttings or buds. In the propagation of grafted grapevines, either the rootstock or the scion may be the source of the virus. Some viruses can spread through seed and pollen. Virus particles may sometimes be detected by serological techniques, but virus diseases in perennial plants such as grapevines are usually diagnosed on the basis of symptoms in the infected host or in inoculated sensitive indicator plants.

Nematodes

Nematodes are small (15–35 × 300–1,000 μm), wormlike, root- or soil-inhabiting organisms. They are round in cross section and have smooth, unsegmented bodies, without appendages. In some species the female swells at maturity, becoming pear-shaped or spheroid. Parasitic nematodes have stylets for piercing host cells during feeding. They may feed only on cells near the root surface and not enter the roots (ectoparasites), or they may enter the roots and feed from within (endoparasites). They may be sedentary or they may migrate slowly in the soil. Nematodes are disseminated by wind, water, and the movement of infested soil, plants, or machinery.

Parasitic Higher Plants

Parasitic higher plants in the families Convolvulaceae (e.g., dodder), Orobanchaceae (e.g., broomrape), and Santalaceae have been observed on grapevines occasionally but are not included in this compendium because none is reported to cause problems in commercial grape production.

Selected References

Agrios, G. N. 1978. Plant Pathology. 2nd ed. Academic Press, New York. 703 pp.

Anderson, H. W. 1956. Diseases of Fruit Crops. McGraw-Hill, New York. 501 pp.

Flaherty, D. L., Jensen, F. L., Kasimatis, A. N., Kido, H., and Moller, W. J., eds. 1981. Grape Pest Management. Publ. 4105. Division of Agricultural Sciences, University of California, Berkeley. 312 pp.

Galet, P. 1977. Les Maladies et les Parasites de la Vigne: Les Champignons et les Virus. Vol. 1. Paysan du Midi, Montpellier, France. 871 pp.

Weaver, R. J. 1976. Grape Growing. John Wiley & Sons, New York. 371 pp.

Winkler, A. J., Cook, J. A., Kliewer, W. M., and Lider, L. A. 1974. General Viticulture. 2nd ed. University of California Press, Berkeley. 710 pp.

(Prepared by R. C. Pearson and A. C. Goheen)

The Family Vitaceae and *Vitis* Speciation

Wild and cultivated grapevines belong to the family Vitaceae, which includes 14 living and two fossil genera and more than a thousand species. The plants are herbaceous or woody vines, with tendrils always arising opposite a leaf. Their inflorescences are generally located in place of a tendril but are rarely axillary or pseudoaxillary. Plants are perfect or unisexually male or female.

The Genus *Vitis*

Grapevines are perennials, annually producing shoots that have tendrils. The more or less branched flower clusters are opposite the leaves. Wild plants are perfect (as are cultured grapevines) or unisexually male or female. The flowers generally have five parts, more rarely four, six, or up to nine parts.

Wild grapevines occur primarily in the Northern Hemisphere, especially in the temperate zones of Asia, North America, Central America, and northwest South America in the Andes chain (Colombia and Venezuela). Cultivated grapevines now exist on five continents wherever the climatic conditions are favorable. In tropical and subtropical areas, grapevines may grow continuously and produce more than one crop per year.

The genus *Vitis* is divided into two sections, *Vitis* (*Euvitis*) and *Muscadinia.*

The section *Vitis* (*Euvitis*) consists of the true grapevines. The canes have an extracambial bark (including the pericyclic fibers and the primary and nonfunctioning secondary phloem) that may shed in strips. The secondary phloem has alternating tangential layers of hard and soft phloem. A diaphragm interrupts the pith at the nodes. Berries ripen more or less evenly on the cluster. Seeds are generally pear-shaped. The basic chromosome number is $n = 19$ or $2n = 38$.

Asiatic species (*V. amurensis, V. davidii, V. armata, V. romanetii, V. piasezkii,* and *V. coignetiae,* among others) are more or less susceptible to phylloxera and to the American grape diseases (black rot, downy mildew, and powdery mildew). *V. amurensis* (which is native to the Amur River in southern Siberia, northern China, and Korea) has been used for breeding cold-tolerant hybrids.

Some American species have played important roles in genetic improvement of grapes around the world. Interspecific American crosses have been used to obtain rootstocks resistant to a variety of soil pests and conditions. American species have been crossed with the European species, *V. vinifera,* to create both rootstocks and fruiting cultivars, including French-American wine grapes, or "French hybrids" (also referred to as "direct producers"), that are more cold-hardy and phylloxera-resistant and less susceptible to fungus diseases than *V. vinifera* cultivars. The discussion that follows is limited to the important species of *Vitis* used in commercial viticulture, as scion cultivars or rootstocks, or in breeding programs.

V. vinifera L.

Civilization spread the Eurasian species *V. vinifera* westward from Asia Minor throughout Europe and eastward throughout Asia. Over the centuries the species has become very diverse; an estimated 5,000 cultivars are in existence today. Although very tolerant of calcareous soil, *V. vinifera* is extremely susceptible to all the American pests and diseases, including phylloxera, downy mildew, powdery mildew, black rot, and Pierce's disease. It is also susceptible to Botrytis bunch rot, anthracnose, Phomopsis cane and leaf spot, Eutypa dieback, and crown gall.

Because of its tolerance to lime, *V. vinifera* is the parent with *V. berlandieri* of several commercial rootstocks, including 41 B, 333 E.M., and Fercal. Because of its fruit quality, *V. vinifera* is in nearly all interspecific hybrids for both wine and table grapes. New intraspecific crosses are continually being introduced in most countries where *V. vinifera* is the predominant commercial species.

V. labrusca L.

Distributed rather rarely in wild areas of the United States east of the Mississippi River, especially in New England, *V. labrusca* and its hybrids account for numerous grape cultivars growing around the world. Cultivars such as Concord, Isabella, Niagara, Noah, Othello, and Campbell Early are grown in rainy regions because of their resistance to fungus diseases.

V. labrusca is moderately resistant to powdery mildew, downy mildew, and winter cold. However, it is susceptible to black rot, Pierce's disease, phylloxera, and lime chlorosis. Rooting and grafting are easily accomplished.

V. riparia Michx.

The North American species *V. riparia* is found most abundantly in southern Canada and from Iowa to states bordering Canada. Although leaves may become covered with phylloxera galls, the roots are very resistant to phylloxera. Cuttings root and graft well, making this species a good genetic source for rootstocks. Its resistance to lime is low.

Gloire de Montpellier, a selection of *V. riparia,* has served as a commercial rootstock for the past century. Interspecific crosses with *V. riparia* have produced some of the most important commercial rootstocks: *V. riparia* × *V. rupestris*—3309 C (Couderc 3309), 101-14 Mgt, Schwarzmann; *V. riparia* × *V. berlandieri*—420 A, 34 E.M., 161-49 C, 5 BB, SO 4, 8 B; *V. riparia* × *V. rupestris* × *V. vinifera*—197-17 Cl; *V. riparia* × *V. rupestris* × *V. candicans*—Solonis, 1616 C, 216-3 Cl.

An early-maturing species, *V. riparia* is resistant to winter cold. Although not resistant to Pierce's disease, *V. riparia* is resistant to downy mildew, powdery mildew, and black rot. It is slightly susceptible to thrips and to Septoria leaf spot in the fall. *V. riparia* has been widely used in breeding direct producers: *V. riparia* × *V. labrusca*—Noah, Clinton; *V. riparia* × *V. labrusca* × *V. vinifera*—Othello; *V. riparia* × *V. vinifera*—Baco noir, Oberlin noir; *V. riparia* × *V. rupestris* × *V. vinifera*—Leon Millot, Marechal Foch.

V. rupestris Scheele

The natural habitat of *V. rupestris* was along the southern Missouri River, in Arkansas, Oklahoma, Texas, Mississippi, Louisiana, Tennessee, and northeastern Mexico. This species is now rare in the wild because of urbanization and livestock grazing. *V. rupestris* is resistant to phylloxera and adapts quite well to calcareous soil. It roots and grafts well and is considered a strong and vigorous rootstock. It has good resistance to downy mildew, powdery mildew, black rot, and Pierce's disease but is somewhat susceptible to anthracnose and Phomopsis cane and leaf spot.

Commercial rootstocks derived from this species include the following: pure *V. rupestris*—St. George; *V. rupestris* × *V. riparia*—see *V. riparia; V. rupestris* × *V. berlandieri*—99 R (Richter 99), 110 R, 1103 P, 140 Ru; *V. rupestris* × *V. vinifera*—AXR 1, 1202 C, 93-5 C. Many of the French-American cultivars bred by the French viticulturists Couderc, Seibel, and Seyve-Villard have *V. rupestris* in their parentage.

V. berlandieri Planch.

V. berlandieri is found in Texas and northern Mexico. It has good resistance to phylloxera and tolerates calcareous soil. Although it grafts well, it is difficult to root. Its lime tolerance, critical for many European growers, has been incorporated into commercial rootstocks through interspecific hybridization. As noted under *V. vinifera* above, several *V. berlandieri* × *V. vinifera* hybrids have been bred for soils containing 25% or more active lime: 41 B, 333 E.M., and Fercal. *V. berlandieri* is also resistant to downy mildew, powdery mildew, black rot, and Pierce's disease. In some years powdery mildew and downy mildew have been observed on the foliage, but not in significant amounts. *V. berlandieri* also was used by Seibel in breeding French-American hybrids.

V. aestivalis Michx.

Widely found in the eastern half of the United States from Wisconsin to Florida, *V. aestivalis* is also known under the names of *V. lincecumii, V. bicolor,* and *V. bourquina* or *V. bourquinana.* It is resistant to downy mildew, powdery mildew, and Pierce's disease. Because of insufficient resistance to phylloxera, intolerance to lime, and poor rooting ability, *V. aestivalis* offers little for rootstock breeding. Because this species is present in Jaeger 70 (*V. rupestris* × *V. lincecumii*), many French-American hybrids, especially those of Seibel, contain a fraction of *V. aestivalis* in their lineage. It has also been used by American breeders as a parent for Pierce's disease-resistant progeny.

V. candicans Engelm.

V. candicans is found in Oklahoma, Texas, Arkansas, southwestern Louisiana, and northern Mexico. *V. doaniana, V. longii,* and *V. champinii* are related species.

V. candicans is resistant to downy mildew, powdery mildew, black rot, Pierce's disease, and root-knot nematodes. Its resistance to phylloxera is mediocre, and it suffers in calcareous soil. However, it performs well in droughty conditions and transmits to its descendants a certain resistance to chloride injury. The *V. candicans* × *V. riparia* × *V. rupestris* hybrid, 1616 C, as well as Salt Creek, Dog Ridge, Harmony, and Freedom, are commercial rootstocks resistant to nematodes. Because of the very unpleasant fruit flavor it imparts to its hybrids, *V. candicans* has not been used in breeding fruiting cultivars.

V. cordifolia Michx.

V. cordifolia is found from Pennsylvania to Florida (rarely in the Gulf states) and in Illinois, Ohio, Tennessee, Missouri, Kansas, Indiana, Arkansas, Oklahoma, and northern Texas. The leaves do not carry phylloxera galls, but the degree of root resistance is unknown. This species is fairly resistant to downy mildew but is susceptible to powdery mildew and grows poorly in calcareous soil. Cuttings are difficult to root.

V. cinerea Engelm.

V. cinerea is found in southern Illinois, southern Indiana, Missouri, Kansas, Oklahoma, and eastern Texas and from northern Mexico to Georgia and South Carolina. It has good resistance to phylloxera, downy mildew, and black rot. However, it grows poorly in calcareous soil and is difficult to root.

Other wild American species of unknown commercial significance include *V. rubra* (syn. *V. palmata*), *V. californica, V. arizonica, V. coriacea* (syn. *V. shuttleworthii*), *V. gigas, V. monticola, V. rufotomentosa,* and *V. caribaea* (syns. *V. indica, V. tiliaefolia*).

The section *Muscadinia* comprises three American species: *V. rotundifolia, V. munsoniana,* and *V. popenoei.* Various cultivars of *V. rotundifolia,* popularly called muscadines, are cultivated in the southeastern United States for fresh fruit and wine. Wild muscadines are found as far north as Virginia.

The morphology of these vines differs from that of the *Vitis* section: there is no diaphragm at the shoot nodes; the tendrils are always simple (not forked); the canes have prominent lenticels; the bark is formed by a periderm just beneath the epidermis and is not shed as a ring bark; the fibers of the secondary phloem are radially arranged; the seeds are boat-shaped, with an oval chalaza surrounded by radiating ridges and furrows; and the basic chromosome number is $n = 20$ or $2n = 40$.

V. rotundifolia Michx.

V. rotundifolia grows from Virginia south to Florida, west along the Gulf states to south-central Texas and Mexico, and also in Tennessee, Arkansas, and perhaps southern Illinois, Missouri, and Kentucky. It is resistant to phylloxera, downy mildew, powdery mildew, the nonmuscadine form of black rot, Pierce's disease, and some nematodes, including *Xiphinema index.* Because of the difficulties of rooting *V. rotundifolia* and of hybridizing it with other *Vitis* species, this source of resistance has only recently been exploited for rootstocks.

Selected References

Bailey, L. H. 1934. The species of grapes peculiar to North America. Gentes Herbarum 3:151–244.

Galet, P. 1956–1964. Cepages et Vignobles de France. 4 vols. Paul Dehan, Montpellier, France. 3,500 pp.

Galet, P. 1979. A Practical Ampelography. L. T. Morton, trans. Cornell University Press, Ithaca, NY. 248 pp.

Planchon, J. E. 1887. Ampelideae Monographiae Phanerogamarum Prodromi. Vol. 5, Part 2. Pages 304–654 in: Monographiae Phanerogamarum. A. L. P. P. De Candolle and C. P. De Candolle, eds. G. Masson, Paris.

Viala, P. 1889. Une mission viticole en Amérique. C. Coulet, Montpellier, France. 387 pp.

(Prepared by P. Galet and L. T. Morton)

Grapevine Structure and Growth Stages

The structure and development of the grapevine are presented with reference to the *Euvitis* section of *Vitis.* Grapevine development is described following a calendar sequence. Some details are described in Table 1 and the legends of accompanying figures. Technical terms are defined in the glossary at the back of this compendium.

Grapevines are perennials grown from rooted hardwood or softwood cuttings (own-rooted) or from grafts of the scion cultivar on a rootstock. Winter dormancy separates the two growing seasons involved in fruit production (Fig. 2).

Reactivation of Buds, Phloem, Cambium, and Roots

In spring, as soon as the soil thaws, sap rises and drips (bleeds) from pruning cuts. When air temperatures reach 5° C, last year's phloem within the cane at the base of each swelling bud resumes its transportation of assimilates to the growing tissues. This reactivation spreads down the cane from each swelling bud and gradually down the trunk into the roots. Soon cells in the cylinder of thin-walled cambium divide to form new xylem on the inside and new phloem toward the outside of the cane.

Xylem (wood) is the water-conducting, food-storing, and supporting tissue of the plant. It is composed of vessels, parenchyma, and fibers. Phloem is the food-conducting and storing tissue. It is composed of sieve elements, companion cells, parenchyma, and fibers. Mycoplasmas and viruses are known to travel in the sieve elements. In viticulture, all the tissues external to the cambium (phloem and periderm) are termed bark. The year-old bark is shed annually (ring bark).

Both the rejuvenated and newly formed root tips absorb water and nutrients from the soil.

The development of overwintering buds into fruit-bearing shoots has been divided into stages (Table 1 and Fig. 3) for convenience in pest control and management practices. During bud swell the hard outer bud scales separate, exposing the hairy outer surfaces of the stipules enveloping the young leaves. The shoot elongates, and the preformed leaves enlarge slowly as the shoot enters a period of slow growth. The growth rate depends primarily on temperature.

Rapid Shoot and Root Growth

As temperature increases and the leaves mature, the shoot grows rapidly. This vegetative growth includes both the expansion of preformed leaves and internodes and the production of new leaves and internodes. The apex of a growing shoot tip alternately produces new leaf and tendril primordia.

V. vinifera and French-American hybrid cultivars have a discontinuous pattern of nonfoliar appendages (Fig. 4). Two nodes with a leaf opposite a cluster or tendril are followed by a node having a leaf only. Cultivars derived from *V. labrusca* may have a cluster or tendril opposite any leaf. With either pattern, clusters are produced at the more basal nodes, that is, those preformed in the overwintering bud. Once the change to tendrils occurs, only tendrils are produced at the younger nodes. Tendrils usually branch once. When the pressure-sensitive tips make contact with an object, the whole tendril coils, bringing the shoot closer to that object.

The leaf consists of a blade, a petiole, and a pair of ephemeral stipules partially encircling the node (Fig. 4). The blade is composed of a palmate framework of veins that is continuous with the vascular bundles of the petiole and the shoot. Leaves are mature when expanded to full size relative to adjacent leaves on the same shoot, when the mesophyll contains intercellular spaces, and when stomata on the abaxial surface function in gas exchange.

The bud in the axil of each leaf is called a lateral bud. This bud grows immediately into a lateral shoot. It bears a scale (prophyll) at its first node. In the axil of this prophyll, the primary bud of year two (the central bud of the compound winter bud) develops. The primary bud bears usually two prophylls, each with its axillary bud (Fig. 5). The older bud, which is nearer the leaf scar, is the secondary bud; the bud farthest from the leaf scar is the tertiary bud. These buds lie in a line roughly parallel to the longitudinal axis of the cane. After winter dormancy, usually only the primary bud becomes the main shoot of year two.

The secondary and tertiary buds of year one often do not

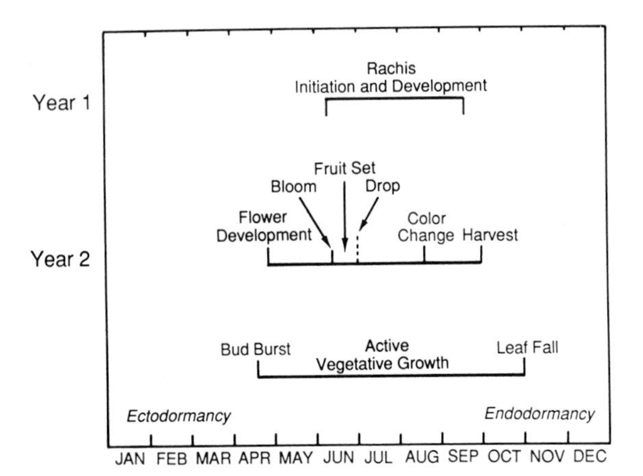

Fig. 2. Chronological relationships of reproductive and vegetative development of Concord grape in New York. (Adapted from Pratt, 1971; used by permission)

TABLE 1. Stages in Shoot Development in the Grapevine[a]

Eichhorn-Lorenz Stages[b]	Baggiolini Stages[c]
01 Winter dormancy: winter bud scales more or less closed	A Winter bud: bud nearly completely covered by two brownish scales
02 Bud swelling: buds expand inside the bud scales	
03 Wool (doeskin stage): brownish wool clearly visible	B Bud swell
05 Bud burst: green shoot first clearly visible	C Green shoot
07 First leaf unfolded and spread away from shoot	D Leaf emergence: tips of leaves visible, bases still protected by wool
09 Two to three leaves unfolded	E Leaves unfolded; first leaves spread away from shoot; internodes visible
12 Five to six leaves unfolded; inflorescences clearly visible	F Four to six leaves unfolded; all inflorescences visible
15 Inflorescence elongating; flowers closely pressed together	G Inflorescences separated and spaced along shoot
17 Inflorescence fully developed; flowers separating	H Flowers separated
19 Beginning of flowering; first caps falling	
21 Early flowering: 25% of caps fallen	
23 Full flowering: 50% of caps fallen	I Flowering
25 Late flowering: 80% of caps fallen	
27 Fruit set: young fruits beginning to swell, remains of flowers lost	J Fruit set
29 Berries small; bunches begin to hang	
31 Berries pea-sized; bunches hang	
33 Beginning of berry touch	
35 Beginning of berry ripening; beginning of loss of green color (*véraison*)	
38 Berries ripe for harvest	
41 After harvest, end of wood maturation	
43 Beginning of leaf fall	
47 End of leaf fall	

[a]Adapted from OEPP/EPPO (1984); used by permission.
[b]Data from Eichhorn and Lorenz (1977). See also Fig. 3.
[c]Data from Baggiolini (1952).

grow into shoots in year two but become latent buds. Shoots derived from latent buds on a trunk, cordon, or arm are called water sprouts and are often mistakenly thought to arise de novo as adventitious shoots. The accumulation of latent buds accounts for the remarkable ability of grapevines to sprout after severe pruning, low-temperature injury, etc.

Roots make their most rapid growth, both in the thickening of permanent roots and in the production of new root tips, during bloom and, to a lesser extent, at harvest.

Maturation

In late summer the main shoots grow at a slower rate. The shoot tip (Fig. 4) produces new nodes, but it never develops bud scales or becomes dormant. Internodal elongation ceases progressively from the base toward the tip of the shoot. If brown periderm (bark) develops, the "mature" shoot after leaf fall is called a cane.

Some summer lateral shoots abscise above the first node, leaving behind a scar and the compound winter bud. Other laterals (called persistent laterals) grow vigorously, often branch, and develop periderm.

Cluster Initiation and Development

Clusters are initiated in the primary and secondary buds. By the time of bud dormancy, the cluster is a branched structure, but no flower parts have developed.

Flower Formation

Flowers are initiated during the spring of year two, between bud swell and bloom. Flower buds just before bloom (anthesis) are covered by the interlocking petals (cap or calyptra) (Fig. 6A). At anthesis the cap separates from the base of the ovary and falls off (Fig. 6B). The stamens spread out (Fig. 6C), and pollen is shed and falls onto the stigma of the pistil. This is

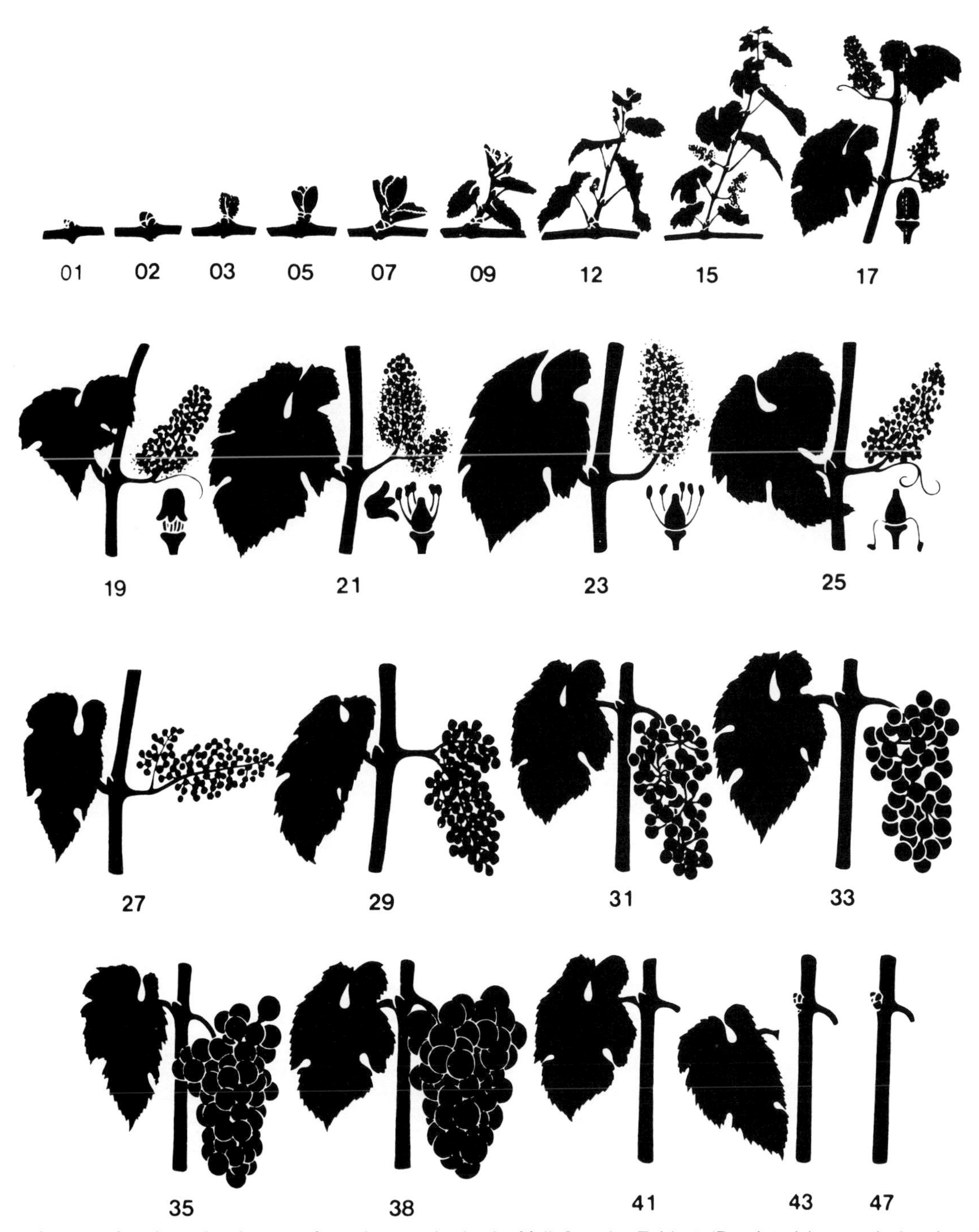

Fig. 3. Stages in grapevine shoot development from dormant bud to leaf fall. See also Table 1. (Reprinted, by permission, from Eichhorn and Lorenz, 1977)

bloom, and it lasts from two to seven days, depending on temperature.

Pollen germinates on the stigma soon after bloom. The pollen tube grows, at a rate depending on temperature, down the center of the style and enters a mature ovule, where fertilization (union of sperm and egg) is effected. Not all ovules are capable of fertilization. This is normal in grape cultivars, but disease or genetic or environmental factors may increase the number of immature or aborted ovules or decrease the germinability of the pollen.

The flowers of cultivated grapes are usually hermaphroditic (perfect). Wild grapes are often dioecious, the female vines having reflexed stamens with nonfunctional pollen and males having reduced, nonfunctional pistils.

Fruit Set

After fertilization the stamens are shed and the ovary begins to enlarge. One to two weeks after bloom, depending on temperature, many small berries fall off. This drop or shatter period is normal in all grape cultivars. The presence of at least one developing berry is necessary for the continued growth in length and girth of the rachis (Fig. 7).

If drop is excessive and the cluster dies or consists of only a few berries, the condition is called *coulure* by the French. Some cultivars are prone to *coulure*. It may result from genetic sterility or from nutritional, environmental, or pathological conditions unfavorable to ovule or pollen maturation.

Another disorder of fruit set, *millerandage* in French, is the production of many small, seedless, "shot" berries. It appears to be related to genetic, nutritional, and environmental factors.

Berry Growth and Maturation

Grape cultivars may be classified into two groups on the basis of the seeded or "seedless" nature of their berries.

Berry growth to maturity takes about 100 days for Concord grapes (seeded) in the Great Lakes region of the United States. After the drop period, the increase in weight of seeded berries is generally divided into three phases: a period of rapid growth until the seed reaches its mature size; a period of slow growth ending with the beginning of loss of green color (*véraison*); and a period of rapid growth ending in maturity, as indicated by color and ratio of soluble solids to acid. The length and distinctness of these periods vary with genetic and environmental factors.

The mature seeded berry is shown in Fig. 8. The epidermis (outer layer of the skin) has a few stomata, which become lenticels as the berry matures. In a given seeded cultivar, mature berry size is related to the number of seeds.

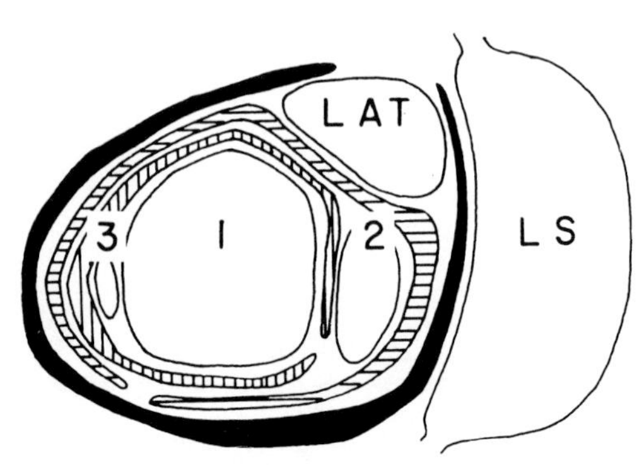

Fig. 5. Transverse section through a compound bud (eye) of Concord grape (×10), showing the relative positions of the leaf scar (LS), lateral shoot scar (LAT), and three dormant buds (1–3): 1, the primary bud, in the axil of the prophyll (solid black) of the lateral shoot; 2, the secondary bud, in the axil of the basal prophyll (horizontally hatched) of the primary shoot; 3, the tertiary bud, in the axil of the next higher prophyll (vertically hatched) of the primary shoot. (Reprinted, by permission, from Pratt, 1974)

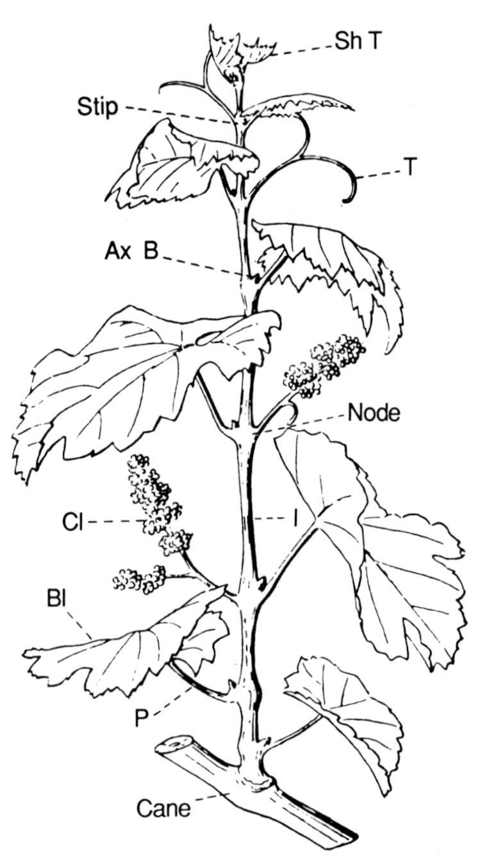

Fig. 4. Shoot of *V. vinifera* at bloom, showing the arrangement of leaves, clusters (Cl), and tendrils (T). The axillary buds (Ax B) develop into summer lateral shoots, each bearing a compound winter bud on the overwintering cane. Bl, blade; I, internode; P, petiole; Sh T, shoot tip; Stip, stipule. (Adapted from von Babo and Mach, 1923; used by permission)

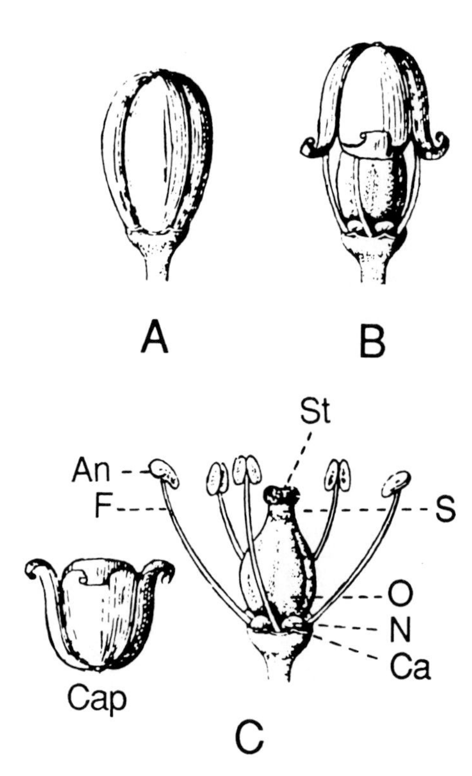

Fig. 6. Stages in the development of a grape flower. **A,** Bud. **B,** Loosening of the cap. **C,** Flower after cap fall. An, anther; Ca, calyx; F, filament; N, nectary; O, ovary; S, style; St, stigma. (Adapted from von Babo and Mach, 1923; used by permission)

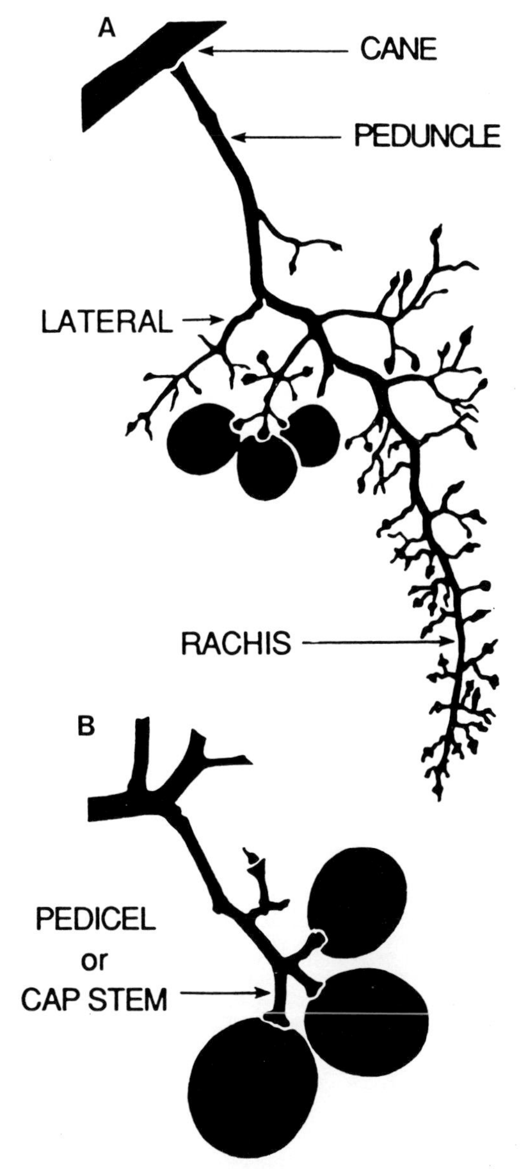

Fig. 7. **A,** Grape cluster with most of the berries removed. **B,** Detail showing berry attachment. (Adapted from Flaherty et al, 1981; used by permission)

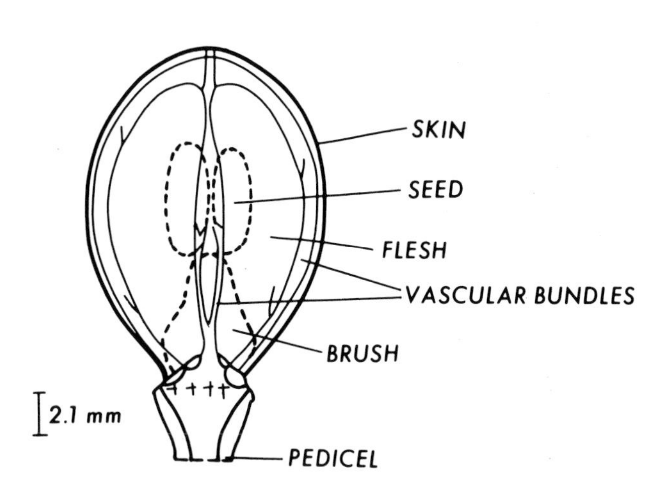

Fig. 8. Longitudinal section of a mature Concord grape berry. When a presenescent berry is picked, the central and lateral vascular strands and adhering flesh tear to form the "brush" or "wet drop." The intact senescent berry or "dry drop" falls off the pedicel at the abscission line (+++). (Adapted from Pratt, 1971; used by permission)

Berries of seedless cultivars grow more uniformly than seeded berries, with a much reduced period of slow growth. They fall into two classes with respect to the time of seed abortion: those in which the ovules do not develop past bloom, like Black Corinth, and those in which the developing seeds abort (seedless table cultivars). It is usually possible to find traces of aborted seeds in the latter class.

Selected References

Baggiolini, M. 1952. Les stades repères dans le développement annuel de la vigne et leur utilisation pratique. Stn. Fed. Essais Agric. (Lausanne) Publ. 12 (MC). 3 pp.

De la Harpe, A. C., Swanepoel, J. J., and Swart, J. P. J. 1982. The anatomy of the genus *Vitis*: An annotated bibliography. S. Afr. J. Enol. Vitic. 3:2, Suppl. 6 pp.

Eichhorn, K. W., and Lorenz, D. H. 1977. Phänologische Entwicklungsstadien der Rebe. Nachrichtenbl. Dtsch. Pflanzenschutzdienstes (Braunschweig) 29:119–120.

Einset, J., and Pratt, C. 1975. Grapes. Pages 130–153 in: Advances in Fruit Breeding. J. Janick and J. N. Moore, eds. Purdue University Press, West Lafayette, IN. 623 pp.

Flaherty, D. L., Jensen, F. L., Kasimatis, A. N., Kido, H., and Moller, W. J., eds. 1981. Grape Pest Management. Publ. 4105. Division of Agricultural Sciences, University of California, Berkeley. 312 pp.

OEPP/EPPO (European and Mediterranean Plant Protection Organization). 1984. EPPO crop growth stage keys. Bull. OEPP/EPPO Bull. 14:295–298.

Pratt, C. 1971. Reproductive anatomy in cultivated grapes—A review. Am. J. Enol. Vitic. 22:92–109.

Pratt, C. 1974. Vegetative anatomy of cultivated grapes—A review. Am. J. Enol. Vitic. 25:131–150.

von Babo, A. F., and Mach, E. 1923. Handbuch Weinbaues und der Kellerwirtschaft. 4th ed., Vol. 1, Part 1. Parey, Berlin. 626 pp.

(Prepared by C. Pratt)

Historical Significance of Diseases in Grape Production

The effects of diseases on grape production are found throughout the records of viticulture. Diseases affect production, harvesting, processing, marketing, and the consumer. They lower quality, reduce yield, and increase the costs of production and harvesting. Diseases debilitate and kill vines and destroy vineyards not only locally but also over large areas and regions and have rendered some land unfit for viticulture. Catastrophic diseases alter crop patterns and may have long-term effects on local and export markets. Some viticulturists emigrated from France to Algeria in the mid-1800s because of epidemics of grape mildews and the root pest phylloxera.

Diseases in general are endemic (native, prevalent, and well-established). Disease development depends on weather, and losses may range up to 5%. In weather favorable to disease, epidemics may occur with losses ranging from 20 to 80%. For example, prolonged periods of wet weather favor Botrytis bunch rot, downy mildew, and other fruit and leaf spot diseases. Phomopsis cane and leaf spot can cause devastating epidemics in protracted periods of wet, cool weather. In contrast, powdery mildew is favored by dry, relatively cool weather.

Some past epidemics illustrate the powers of pathogens and the consequences of diseases. Some pathogens common to the eastern United States caused devastating losses when introduced into Europe and the western United States. In the mid-1800s, the powdery mildew fungus entered Western Europe and spread throughout the continent. The pathogen was named *Oidium tuckeri* Berk. for the gardener who first noted it in Margate, England, in 1845. The disease was first noted in France in 1847 and caused havoc to vines and produced fruit that made "foul wine." Mr. Tucker recognized the similarity between the grape pathogen and one on peach, for which a mixture of sulfur, lime, and water gave control. Losses in France had at times reached 80% by 1854, when dusting with sulfur was perfected as a control.

The American grape root pest phylloxera was found in southwestern France about 1865 and spread throughout Europe. Losses from the pest were catastrophic. The effects of the feeding of the pest on grape roots are similar to those of a plant disease.

In an effort to develop rootstocks resistant to phylloxera, American *Vitis* species were introduced into Europe for use as breeding material. Inadvertently, pathogens of some grape diseases were introduced with them. The downy mildew fungus was first noted in southwestern France about 1878 and by 1882 had spread to all of France and soon after to the rest of the continent. The disease was highly destructive to vines and fruit, but annual losses varied. In 1885, P. M. A. Millardet first used copper sulfate, lime, and water (Bordeaux mixture) to control downy mildew near Bordeaux, France. This was a historic event, for the fungicide was subsequently used in the control of many fungal and bacterial diseases, and it remained the most important fungicide in the world for over 50 years.

Powdery mildew and phylloxera developed early in the burgeoning viticulture industry in California, but not downy mildew, because the warm, dry weather is unsuitable for disease development. However, a disease known as the California vine disease was first noted near Anaheim about 1884 in a region where vines had been cultured on mission grounds for some 300 years with no records of serious diseases. The mysterious disease was observed in Napa Valley in 1887 and later in other locations. By 1906 the disease had destroyed approximately 16,000 ha of vineyards. In the 1930s during another epidemic, the disease was renamed Pierce's disease after the scientist who studied it extensively in southern California. In the 1970s the pathogen of the disease was shown to be a fastidious vascular bacterium. The disease limits the culture of European and many American cultivars in southern California. The pathogen is apparently a long-time resident of certain areas and also limits the culture of European grapes in parts of the Gulf Coastal Plain of the United States, Mexico, and Central America.

Following the epidemics of the mildews and phylloxera, and where vineyards were reestablished on "resistant" rootstocks, a new insidious, deforming, and degenerating disease developed. Known in different countries as *roncet, court-noué, panachure,* and *Reisigkrankheit,* the disease had a developmental period of 12–15 years from planting, after which production declined rapidly and vines degenerated. Reclaiming soils required approximately 10 years of fallow from vines. By the mid-1930s the disease was known to be transmitted by grafting and to be soilborne, and it was named, in places, infectious degeneration. In the 1950s it was shown to be caused by the grapevine fanleaf virus and vectored by soilborne dagger nematodes. Replanting vineyards with virus-free planting stock in vector-free soils has increased production by 50–100%.

Although grape pathogens are dynamic and their explosive powers are awesome, through research and development they have been brought under control.

Selected References

Gardner, M. W., and Hewitt, W. B. 1974. Pierce's Disease of the Grapevine: The Anaheim Disease and the California Vine Disease. Departments of Plant Pathology, University of California, Berkeley and Davis. 225 pp.

Hewitt, W. B., Goheen, A. C., Raski, J. D., and Gooding, G. V., Jr. 1962. Studies of virus diseases of the grape in California. Vitis 3:57–83.

Large, E. D. 1940. The Advances of the Fungi. Henry Holt and Co., New York. 488 pp.

Millardet, P. M. A. 1885. Traitement du Mildiou et du Rot. J. Agric. Prat. 2:513–516. Pages 7–11 in: The Discovery of Bordeaux Mixture. F. J. Schneiderhan, trans. Phytopathol. Classics 3 (1933). American Phytopathological Society, St. Paul, MN. 25 pp.

Parris, G. K. 1968. A Chronology of Plant Pathology. Johnson and Sons, Starkville, MS. 167 pp.

(Prepared by W. B. Hewitt)

Part I. Diseases Caused by Biotic Factors

Fruit and Foliar Diseases Caused by Fungi

Powdery Mildew

The powdery mildew fungus (also called oidium) was first described in North America by Schweinitz in 1834. The disease caused minor damage on native American grapes and did not gain notoriety until 1845, when it was first observed in England (see Introduction—Historical Significance of Diseases in Grape Production). Today this disease can be found in most grape-growing areas of the world, including the tropics. Uncontrolled, powdery mildew reduces vine growth and yield and affects quality and winterhardiness. Only members of the Vitaceae are susceptible to the causal fungus.

Symptoms

The powdery mildew fungus can infect all green tissues of the grapevine. The fungus penetrates only the epidermal cells, sending haustoria into them to absorb nutrients. Although haustoria are found only in epidermal cells, neighboring noninvaded cells may become necrotic. The presence of mycelia with conidiophores and conidia on the surface of the host tissue gives it a whitish gray, dusty or powdery appearance (Plate 1). Both surfaces of leaves of any age are susceptible to infection. Occasionally, the upper surface of infected leaves exhibits chlorotic or shiny spots that resemble the "oil spot" symptoms of downy mildew. Young, expanding leaves that are infected become distorted and stunted (Plate 2).

Petioles and cluster stems are susceptible to infection throughout the growing season; once infected, they become brittle and may break as the season progresses. When green shoots are infected, the affected tissues appear dark brown to black in feathery patches (Plate 3), which later appear reddish brown on the surface of dormant canes. Only remnants of collapsed hyphal fragments can be found at this stage of development.

Cluster infection before or shortly after bloom may result in poor fruit set and considerable crop loss. Berries are susceptible to infection until their sugar content reaches about 8%, although established infections continue to produce spores until the berries contain 15% sugar.

If berries are infected before they attain full size, the epidermal cells are killed, and growth of the epidermis is thus prevented. As the pulp continues to expand, the berry splits from internal pressure. Split berries either dry up or rot, frequently becoming infected by *Botrytis cinerea.* Berries of nonwhite cultivars that are infected as they begin to ripen fail to color properly and have a blotchy appearance at harvest (Plate 4). A netlike pattern of scar tissue (Plate 5) may be observed on the surface of infected berries. Such fruit is unmarketable as fresh fruit, and wines made from it may have off-flavors.

In most viticultural regions, the fungus produces its sexual structures, black spherical bodies called cleistothecia (Plate 6), on the surface of infected leaves, shoots, and clusters during the latter part of the growing season.

Causal Organism

Uncinula necator (Schw.) Burr. (syns. *Erysiphe necator* Schw., *E. tuckeri* Berk., *U. americana* Howe, *U. spiralis* Berk. & Curt., *U. subfusca* Berk. & Curt.; anamorph *Oidium tuckeri* Berk.), the fungus that causes powdery mildew, is an obligate parasite on the Vitaceae genera *Ampelopsis, Cissus, Parthenocissus,* and *Vitis*. The superficial but semipersistent, septate, hyaline hyphae (4–5 μm in diameter) develop characteristic multilobed appressoria from which penetration pegs are formed. After penetration of the cuticle and cell wall, a globose haustorium is formed within the epidermal cell.

Multiseptate conidiophores (10–400 μm long) form perpendicularly on the prostrate hyphae at frequent intervals. Conidia are hyaline and cylindro-ovoid, measure 27–47 × 14–21 μm, and accumulate in chains (Plate 7). The oldest conidium is at the distal end of the chain. Under field conditions the chains are rather short, with three to five conidia.

Cleistothecia, formed after fusion of hyphae of opposite mating types, are globose (84–105 μm in diameter) and may be found on the surface of all infected parts of the host. Cleistothecia have long, flexuous, multiseptate appendages that have a characteristic crook at the apex when mature. Cleistothecia change from white to yellow to dark brown as they mature (Plate 6). They contain four to six asci (rarely more) that are ovate to subglobose and measure 50–60 × 25–40 μm. Asci contain four to seven (most commonly four at maturity) ovate to ellipsoid, hyaline ascospores measuring 15–25 × 10–14 μm (Fig. 9). Similar to conidia, viable ascospores germinate with one or more short germ tubes, each quickly forming a multilobed appressorium.

Disease Cycle and Epidemiology

U. necator may overwinter as hyphae inside dormant buds of the grapevine, as cleistothecia on the surface of the vine, or both (Fig. 10). In greenhouses and in tropical climates, mycelia and conidia may survive from one season to the next on green tissue remaining on the vine.

Developing buds are infected during the growing season. The fungus grows into the bud, where it remains in a dormant state on the inner bud scales until the following season. Shortly after budbreak, the fungus is reactivated and covers the emergent shoot with white mycelium (Plate 8). Conidia are produced abundantly on these infected shoots (called flag shoots) and are readily disseminated by wind to neighboring vines.

In viticultural regions where cleistothecia are an important source of primary inoculum, the first infections may be observed as individual colonies on the surface of leaves growing in close proximity to bark. Shoots growing near bark are frequently infected first, presumably because they are close to cleistothecia that were trapped in bark crevices after being washed there from leaves, canes, and cluster stems during autumn rains. In spring, the cleistothecia split when wetted by rain, and the ascospores are forcibly discharged. Ascospores germinate and infect green tissue, resulting in colonies that produce conidia for secondary spread.

The effects of environmental factors such as temperature, moisture, and light on the survival and germination of conidia and on colony development have been studied extensively.

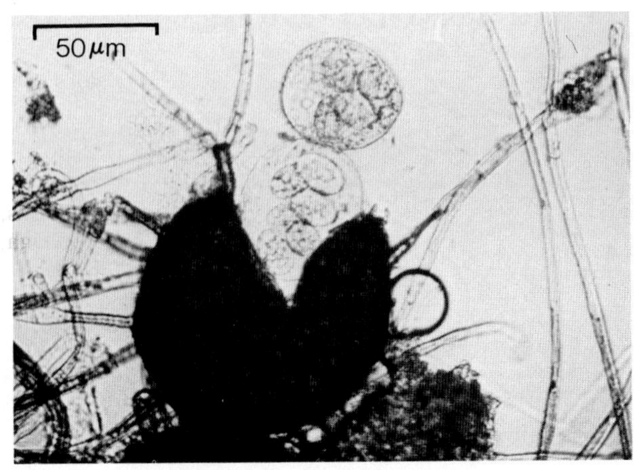

Fig. 9. Cleistothecium of *Uncinula necator* with asci containing ascospores. (Courtesy R. C. Pearson)

Temperature appears to be the major limiting environmental parameter for the development of the fungus. Temperatures of 20–27° C are optimal for infection and disease development, although fungal growth can occur from 6 to 32° C. Temperatures above 35° C inhibit germination of conidia, and above 40° C they are killed. At 25° C, conidia germinate in approximately 5 hr. The time from inoculation to sporulation at 23–30° C can be as short as five to six days, whereas at 7° C, more than 32 days are required. Mildew colonies are reported to be killed after exposure to 36° C for 10 hr or 39° C for 6 hr. Temperature and moisture requirements of ascospores are unknown.

Free water often results in poor and abnormal germination of conidia as well as bursting of conidia, presumably because of excessive turgor pressure. Rainfall can be detrimental to disease development by removing conidia and disrupting mycelium. Atmospheric moisture in the range of 40–100% relative humidity is sufficient for germination of conidia and infection, although germination has also been reported at less than 20% relative humidity. Humidity appears to have a greater effect on sporulation than on germination. For example, two, three, and four to five conidia have been reported to form during a 24-hr period at 30–40, 60–70, and 90–100% relative humidity, respectively.

Low, diffuse light favors disease development. In fact, bright sunlight is reported to inhibit germination of conidia; in one study, germination of conidia was 47% in diffuse light but only 16% in sunlight.

Control

Control of powdery mildew in commercial vineyards is generally based on the use of fungicides. Sulfur was the first effective fungicide used to control this disease and, because of its efficacy (both preventive and curative) and low cost, it is still the most widely used fungicide for this purpose. Sulfur is commonly applied as a dust or as a wettable powder. In dry climates sulfur dust is preferred, whereas in regions where rainfall is plentiful during the growing season, wettable powder

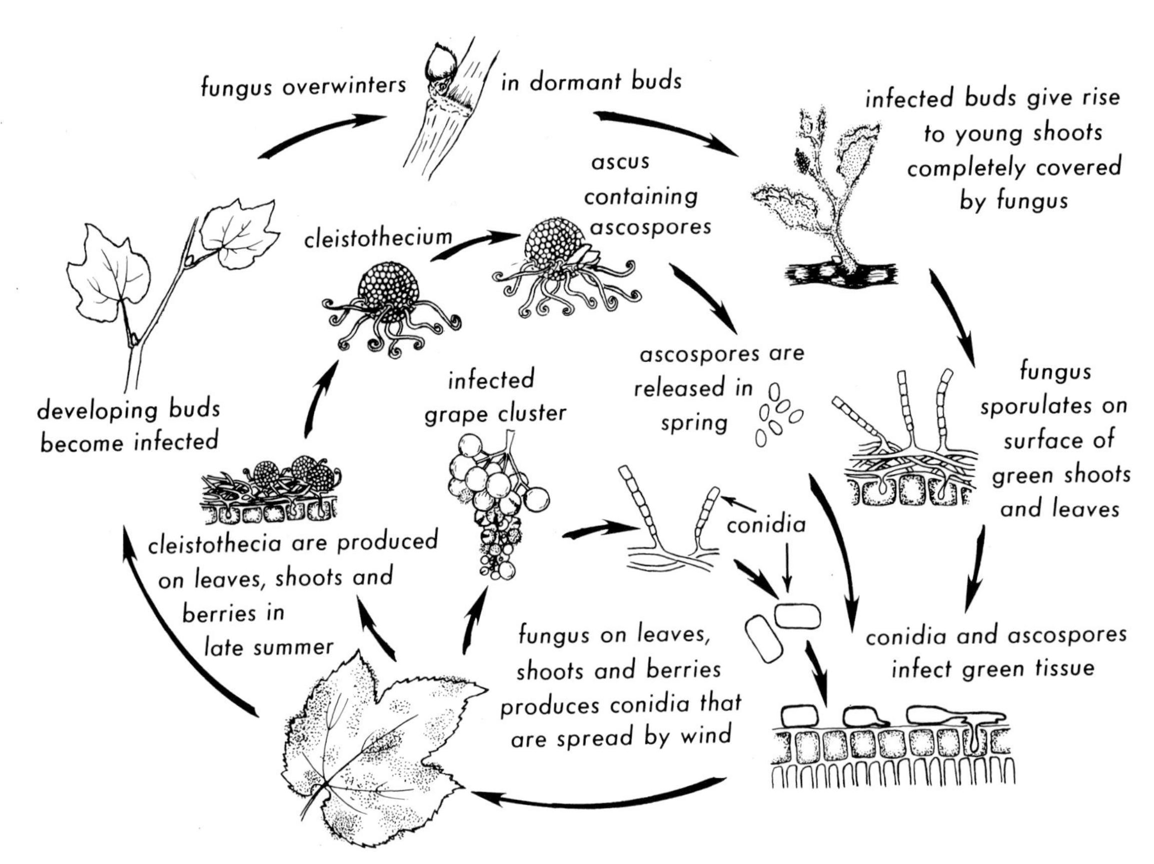

Fig. 10. Disease cycle of powdery mildew. (Drawing by R. Sticht)

or flowable formulations are preferred for their retentive qualities.

Much of the fungicidal activity of sulfur is associated with its vapor phase. The production of vapors and their effectiveness depend on the type of sulfur as well as on environmental factors, primarily temperature. The optimal temperature range for sulfur activity is 25–30° C, and the fungicide may not be effective below 18° C. Above 30° C the risk of phytotoxicity increases greatly, and applications at 35° C or higher are not recommended. Sulfur is less active in humid air than in dry air.

Copper formulations and several organic fungicides, such as dinocap, benomyl, and compounds belonging to the sterol biosynthesis inhibitor group (e.g., triadimefon), are also used commercially to control powdery mildew, although not as extensively as sulfur. The organic fungicides maintain activity over a wider temperature range than sulfur and, with the exception of dinocap, exhibit less phytotoxicity.

Cultural practices may reduce the severity of disease and can increase the effectiveness of chemical control. Planting in sites with good air circulation and sun exposure and orienting rows to take advantage of these factors are helpful. The use of training systems that allow good air movement through the canopy and prevent excess shading is also beneficial. An open canopy not only maintains a microclimate less favorable for disease development but also allows better penetration of fungicide.

Vitis species differ greatly in susceptibility to powdery mildew. *V. vinifera* and Asiatic species such as *V. betulifolia, V. pubescens, V. davidii, V. pagnucii,* and *V. piasezkii* are highly susceptible. By comparison, American species such as *V. aestivalis, V. berlandieri, V. cinerea, V. labrusca, V. riparia,* and *V. rupestris* are much less susceptible. Grape breeders have crossed *V. vinifera* with various combinations of American species to produce hybrids with varying levels of resistance.

To date, biological control has not been applied to *U. necator*. The most commonly reported mycoparasites are *Ampelomyces quisqualis* Ces. (syn. *Cicinnobolus cesatii* De Bary) and *Tilletiopsis* sp. Although use of these fungi may have application in the controlled environment of a greenhouse, they have not been used to control grape powdery mildew commercially under field conditions.

Selected References

Boubals, D. 1961. Étude des causes de la résistance des Vitacées à l'oïdium de la Vigne—*Uncinula necator* (Schw.) Burr.—et de leur mode de transmission héréditaire. Ann. Amelior. Plant. 11:401–500.

Bulit, J., and Lafon, R. 1978. Powdery mildew of the vine. Pages 525–548 in: The Powdery Mildews. D. M. Spencer, ed. Academic Press, New York. 565 pp.

Delp, C. J. 1954. Effect of temperature and humidity on the grape powdery mildew fungus. Phytopathology 44:615–626.

Kapoor, J. N. 1967. *Uncinula necator*. Descriptions of Pathogenic Fungi and Bacteria, No. 160. Commonwealth Mycological Institute, Kew, Surrey, England.

Lafon, R. 1982. Faire face à l'oidium. Vititechnique 57:10–14.

Pearson, R. C., and Gadoury, D. M. 1987. Cleistothecia, the source of primary inoculum for grape powdery mildew in New York. Phytopathology 77:1509–1514.

Pearson, R. C., and Gärtel, W. 1985. Occurrence of hyphae of *Uncinula necator* in buds of grapevine. Plant Dis. 69:149–151.

Pool, R. M., Pearson, R. C., Welser, M. J., Lakso, A. N., and Seem, R. C. 1984. Influence of powdery mildew on yield and growth of Rosette grapevines. Plant Dis. 68:590–593.

Sall, M. A. 1980. Epidemiology of grape powdery mildew: A model. Phytopathology 70:338–342.

(Prepared by R. C. Pearson)

Downy Mildew

Grape downy mildew occurs in regions where it is warm and wet during the vegetative growth of the vine (e.g., Europe, South Africa, Argentina, Brazil, eastern North America, eastern Australia, New Zealand, China, and Japan). The absence of rainfall in spring and summer limits the spread of the disease in certain areas (e.g., Afghanistan, California, and Chile), as does insufficient warmth during the spring in northern vineyards.

Cultivars within *V. vinifera* are highly susceptible to downy mildew. *V. aestivalis* and *V. labrusca* are less susceptible, and *V. cordifolia, V. rupestris*, and *V. rotundifolia* are relatively resistant.

Symptoms

The causal fungus attacks all green parts of the vine, particularly the leaves. Depending on incubation period and leaf age, lesions are yellowish and oily (Plate 9) or angular, yellow to reddish brown, and limited by the veins (Plate 10). Sporulation of the fungus—a delicate, dense, white, cottony growth—characteristically occurs on the lower leaf surface (Plate 11). Leaf infection is most important as a source of inoculum for berry infection and as overwintering inoculum. Severely infected leaves generally drop. Such defoliation reduces sugar accumulation in fruit and decreases hardiness of overwintering buds.

Infected shoot tips thicken, curl ("shepherd's crook"), and become white with sporulation (Plate 12); they eventually turn brown and die. Similar symptoms are seen on petioles, tendrils, and young inflorescences, which, if attacked early enough, ultimately turn brown, dry up, and drop.

The young berries are highly susceptible, appearing grayish when infected (gray rot) and covered with a downy felt of fungus sporulation (Plates 13 and 14). Although berries become less susceptible as they mature, infection of the rachis can spread into older berries (Plate 15) (brown rot, without sporulation). Infected older berries of white cultivars may turn dull gray-green, while those of black cultivars turn pinkish red. Infected berries remain firm compared to healthy berries, which soften as they ripen. These infected berries drop easily, leaving a dry stem scar. Portions of the rachis or the entire cluster also may drop.

Causal Organism

Plasmopara viticola (Berk. & Curt.) Berl. & de Toni, the cause of downy mildew, is an obligate parasite. It develops intercellularly within the parasitized tissues of the vine in the form of tubular, coenocytic hyphae 8–10 μm in diameter, bearing globular haustoria that are 4–10 μm in diameter. The haustoria enter the host cell by invaginating the cellular membrane in which they are ensheathed.

Asexual reproduction occurs by the formation of sporangia, which are ellipsoid and hyaline and measure 14×11 μm. Sporangia are borne on treelike sporangiophores (140–250 μm long) (Fig. 11). Each sporangium gives rise to one to 10 biflagellate zoospores measuring $6–8 \times 4–5$ μm. The zoospores escape from the side of the sporangium opposite its point of attachment, either through an opening in a papilla or by directly perforating the wall. Zoospores are mainly uninucleate. Protoplasmic fusions between hyphae originating from different zoospores may occur inside parasitized tissues and give rise to heterokaryotic mycelium.

Sexual reproduction begins early in the summer and occurs by the fusion of an antheridium and an oogonium derived from the terminal expansion of hyphae. An oospore (20–120 μm in diameter) forms and is enveloped by two membranes and covered by the wrinkled wall of the oogonium. Oospores form in leaves or possibly throughout the parasitized organ. The following spring, oospores germinate in free water, giving rise to one or occasionally two slender germ tubes, 2–3 μm in diameter and of variable length. The germ tubes terminate in a pyriform sporangium (28×36 μm), which produces 30–56 zoospores.

SANDOZ

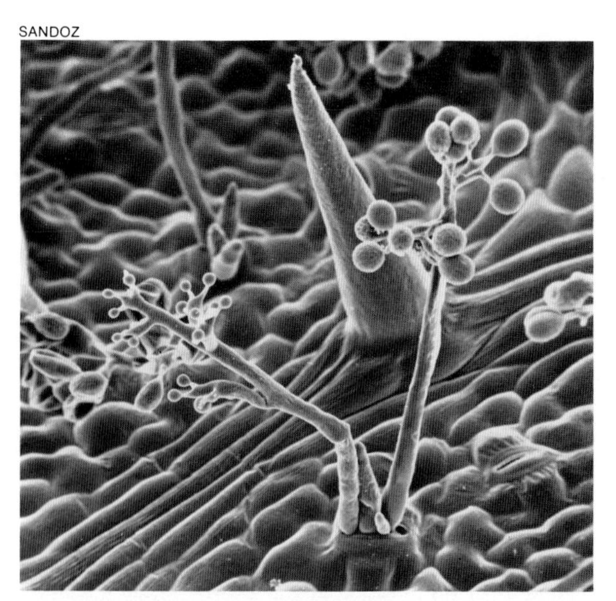

Fig. 11. Sporangia of *Plasmopara viticola* on treelike sporangiophores emerge through stomata on the underside of a leaf. (Micrograph by R. Guggenheim; reprinted, by permission, from "The new orazolidinone class of systemic fungicides" by R. Sandmeier, Produits Sandoz, Rueil-Malmaison Cedex, France)

Disease Cycle and Epidemiology

P. viticola overwinters mainly as oospores in fallen leaves, although it can survive as mycelium in buds and in persistent leaves, the latter in regions with mild winters (Fig. 12). Oospores survive best in the surface layers of moist soil; survival is little affected by temperature. Oospores germinate in water in spring as soon as temperatures reach 11°C to produce a sporangium from which primary dispersal of zoospores occurs by rain-splash.

Sporangiophores and sporangia are produced only through the stomata of infected organs, a process that requires 95–100% relative humidity and at least 4 hr of darkness. The optimal temperature for sporulation is 18–22°C. The sporangia are detached from sporangiophores by the dissolution of a cross-wall of callose, moisture again being required. The sporangia are dispersed by wind to leaves, where they germinate in free water (optimal temperature 22–25°C) to release zoospores.

Zoospores swim to and encyst near stomata, which are entered by germ tubes from germinating cysts. Because the fungus penetrates the host exclusively via the stomata, only those plant structures with functional stomata are susceptible to infection. Under optimal conditions, the time from germination to penetration is less than 90 min.

Because sporangia are usually formed at night and are inactivated by several hours' exposure to sunlight, infection generally occurs in the morning. The time from infection to the

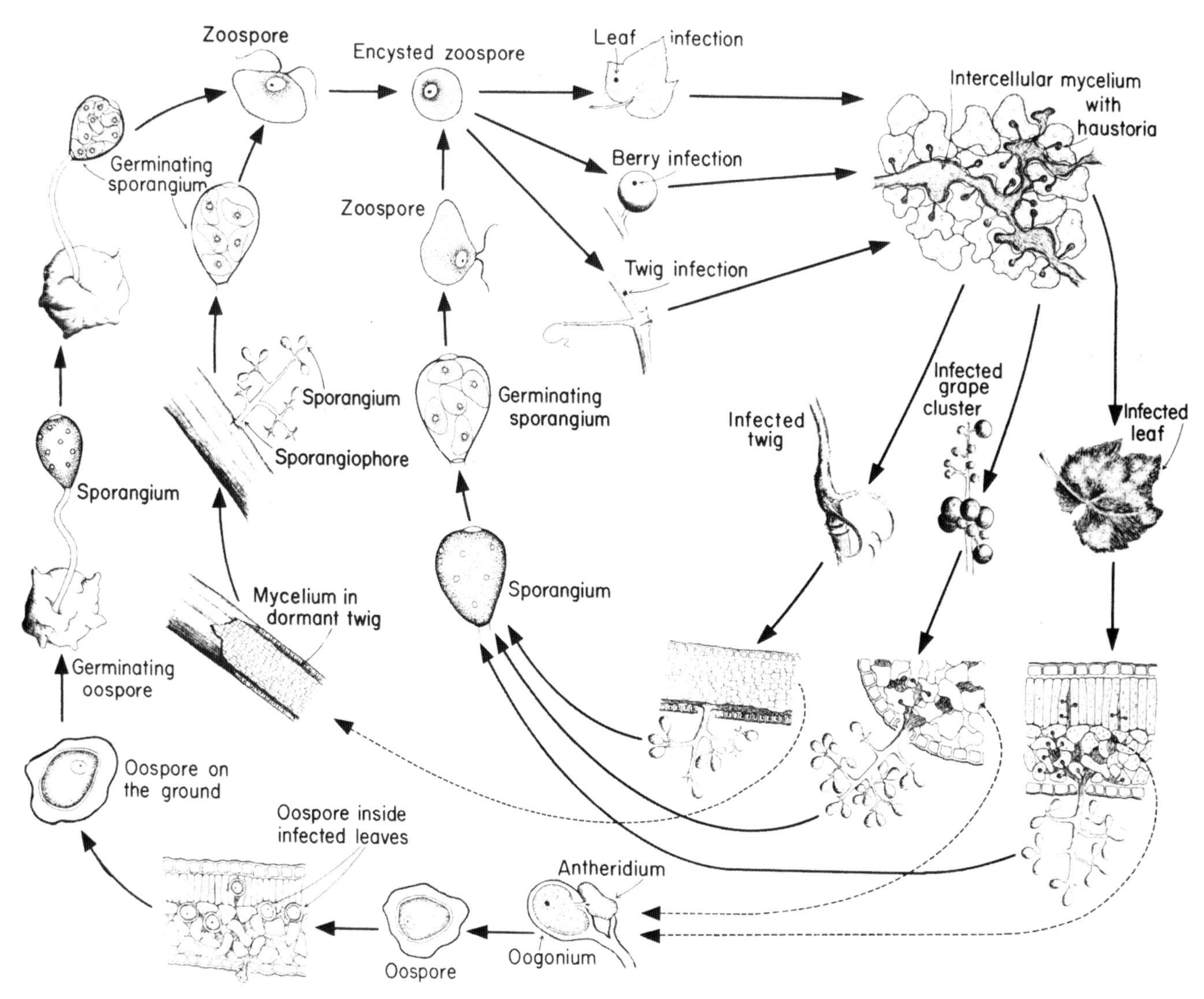

Fig. 12. Disease cycle of downy mildew. (Reprinted, by permission, from G. N. Agrios, Plant Pathology, 2nd ed., 1978, Academic Press, New York)

first appearance of symptoms is approximately four days, depending on leaf age, cultivar, temperature, and humidity.

Downy mildew is favored by all factors that increase the moisture content of soil, air, and host plant. Therefore, rain is the principal factor promoting epidemics. Temperature plays a less important role by retarding or accelerating the development of the disease. The optimal temperature for development of the fungus is about 25° C, the extremes ranging from 10 to 30° C.

The most serious epidemics of downy mildew occur when a wet winter is followed by a wet spring and a warm summer with intermittent rainstorms every 8–15 days. These conditions ensure the survival of oospores and their abundant germination in spring as well as permit the development of the disease and its spread within the vineyard. Successive periods of rain stimulate the production of young, susceptible shoots with functional stomata (preparatory rains) and ensure their infection (infection rains).

Control

Preventive management practices for downy mildew consist of draining soils, reducing the sources of overwintering inoculum, and pruning out the ends of infected shoots. However, because none of these measures is practical or sufficient in vineyards susceptible to downy mildew, chemical control must inevitably be used.

Fungicides are the most important control measure on susceptible cultivars grown in regions with high disease pressure. Nonsystemic surface chemicals (cupric salts, dithiocarbamates, and phthalimides) are efficacious only as preventive treatments. They are fungitoxic at several cellular sites in fungi, and *P. viticola* has not become resistant to them. Only the treated organs are protected, generally for seven to 10 days.

Cymoxanil is a nonsystemic penetrating fungicide specific to mildew. It penetrates the treated organs and synergistically increases the efficacy of nonsystemic surface fungicides combined with it. However, the principal advantage of cymoxanil is its capacity to act curatively during the two or three days following an infection period.

Two classes of systemic fungicides are active against the downy mildew fungus, fosetyl aluminum and the phenylamides (or anti-oomycete anilides). These products penetrate the plant and have three principal advantages: the active substance is not removed by rainfall, treatment is curative, and vegetation formed after treatment is protected. The interval between applications can be 14 days.

The phenylamides (benalaxyl, metalaxyl, ofurace, oxadixyl) are very effective but are specific to *P. viticola,* and resistant strains have developed since 1981 in France, South Africa, Switzerland, and Uruguay. Resistant strains are less competitive than sensitive strains, so the use of phenylamides in combination with at least one nonsystemic fungicide and limited to only two or three applications per year is recommended.

Selected References

Blaeser, M., and Weltzien, H. C. 1977. Untersuchungen über die Infektion von Weinreben mit *Plasmopara viticola* in Abhängigkeit von der Blattnässedauer. Meded. Fac. Landbouwwet. Rijksuniv. Gent 42:967–976.

Blaeser, M., and Weltzien, H. C. 1978. Die Bedeutung von Sporangienbildung, -ausbreitung und-keimung für die Epidemiebildung von *Plasmopara viticola.* Z. Pflanzenkr. Pflanzenschutz 85:155–161.

Lafon, R. 1985. Les fongicides viticoles. Pages 191–198 in: Fungicides for Crop Protection. Vol. 1. I. M. Smith, ed. Monogr. 31. British Crop Protection Council, Croydon, England. 504 pp.

Lafon, R., and Bulit, J. 1981. Downy mildew of the vine. Pages 601–614 in: The Downy Mildews. Academic Press, New York. D. M. Spencer, ed. 636 pp.

Langcake, P., and Lovell, A. 1980. Light and electron microscopical studies of the infection of *Vitis* spp. by *Plasmopara viticola,* the downy mildew pathogen. Vitis 19:321–337.

Leroux, P., and Clerjeau, M. 1985. Resistance of *Botrytis cinerea* Pers. and *Plasmopara viticola* (Berk. & Curt.) Berl. and de Toni to fungicides in the French vineyards. Crop Prot. 4:137–160.

(Prepared by R. Lafon and M. Clerjeau)

Botrytis Bunch Rot and Blight

Botrytis bunch rot or gray mold exists in all vineyards in the world. It was considered a secondary disease for a long time but became increasingly important in Europe after the phylloxera epidemic and the reconstitution of vineyards by grafting. Temperate or cold, damp climates favor this disease.

Botrytis bunch rot seriously reduces quality and quantity of the crop. The reduction in yield may be associated with the premature drop of bunches from stalk rot or with the loss of juice and the desiccation of berries. In table grape production, loss of fruit quality in the field, in storage, or during transit can be substantial. In wine production, the most serious damage is qualitative, from the modified chemical composition of diseased berries. The fungus converts simple sugars (glucose and fructose) to glycerol and gluconic acid and produces enzymes that catalyze the oxidation of phenolic compounds. It also secretes polysaccharides such as β-glucan, which hinder the clarification of wine. Wines produced from rotten grapes have off-flavors and are fragile and sensitive to oxidation and bacterial contamination, making them unsuitable for aging.

However, in certain cultivars and especially under certain climatic conditions in the fall, Botrytis infection of grape clusters takes on a particular form known as "noble rot." This rot is beneficial and contributes to the production of exceptional sweet white wines, the most prestigious of which are the Tokays of Hungary, the Sauternes of France, and the German wines known as *Auslese, Beerenauslese,* and *Trockenbeerenauslese.*

Symptoms

In early spring, buds and young shoots may be infected, turn brown, and dry out. At the end of spring and before bloom, large, irregular, reddish brown, necrotic patches appear on a few leaves of a vine and are often localized on the edge of the lamina (Plate 16).

Before capfall (bloom), the fungus may invade inflorescences, which rot or dry out and fall off (Plate 17). At the end of bloom, *Botrytis* frequently develops on the withered calyptras, stamens, and aborted berries attached to or trapped in the clusters. From these sites it attacks the pedicel or the rachis, forming small patches that are brown at first and then turn black. Toward the end of summer, these lesions completely surround the pedicel or rachis (compare with stem necrosis), and portions of the cluster below the necrotic area wither and drop off (Plate 18).

From *véraison* (ripening) onward, the grapes are infected directly through the epidermis or through wounds. The mold progressively invades the entire cluster. Rot develops rapidly in compact clusters where maturing berries are compressed together (Plate 19). Infected white grapes turn brown, and black grapes become reddish. During dry weather, infected berries dry out; in wet weather, they tend to burst, and a brownish gray mold forms on the surface. Table grapes in cold storage often develop a wet rot of the rachis, which becomes covered by a mycelial mat (Plate 20), sometimes with sporulation. The infected berries develop circular brown lesions that gradually cover the whole fruit; this condition of the epidermis is known as "slip-skin."

In Europe, poorly hardened canes may become infected late in the season and show a bleaching of the bark, combined with the development of black sclerotia or grayish, sporulating patches of mycelium.

Newly grafted grapevine cuttings incubated in callusing

boxes at temperatures of about 30° C and high humidity may be infected and destroyed by rapid growth of the mycelium of *Botrytis.* The fungus also develops under the film of paraffin used to seal the graft union of grafted vines and thereby inhibits development of the union.

Causal Organism

The causal fungus is *Botryotinia fuckeliana* (de Bary) Whetzel, of which only the conidial form, *Botrytis cinerea* Pers., is generally observed in vineyards. The mycelium of *B. cinerea* is composed of brownish olive, septate hyphae, which are cylindrical or slightly swollen at the septa. The hyphae vary in diameter (11–23 μm) according to the conditions of development. Anastomoses between hyphae are often noted.

Conidiophores (1–3 mm long) are stout, dark, slender, and branched, with enlarged apical cells bearing clusters of conidia on short sterigmata (Fig. 13). Conidia (10–12 × 8–10 μm) are ovoid or globose, smooth, one-celled, slightly ash-colored, and gray in mass.

Under adverse conditions the fungus produces sclerotia (2–4 × 1–3 mm), which are dark, discoid, and firmly attached to the substrate. They consist of a medulla and a dark cortical layer of cells. Sclerotia can germinate at temperatures from 3 to 27° C by conidiophore production.

B. cinerea may also produce microconidia (phialospores). The phialides usually arise freely from single hyphal cells on the old aerial mycelium. The microconidia (2–3 μm in diameter) are hyaline and one-celled, are formed in chains, and are embedded in mucilage. Their sole function is the spermatization of sclerotia, leading to the formation of apothecia.

Sclerotia can germinate to form the apothecia of *Botryotinia fuckeliana,* but apothecia are rarely found in vineyards. Apothecia are cupulate, stalked, and brownish, with a stipe about 4–5 mm long. Ascospores (7 × 5.5 μm) are hyaline, one-celled, ovoid-ellipsoid, and smooth.

Disease Cycle and Epidemiology

B. cinerea is not specific to grapevines. It attacks many cultivated and wild plants and can live as a saprophyte on necrotic, senescent, or dead tissue.

The pathogen overwinters as sclerotia (in Europe), formed in the autumn on the canes (sometimes on mummified grapes), and also as mycelium on the bark and in dormant buds. In spring, the sclerotia and the mycelium produce conidia, which are probably the source of inoculum for prebloom infection of leaves and young clusters. Conidia are disseminated by rain and wind and are considerably more numerous after *véraison.*

Conidia germinate at temperatures between 1 and 30° C (18° C is optimal). In water, germination is stimulated by exogenous nutrients from pollen or leaf exudates. In the absence of water, germination occurs if the relative humidity is at least 90%. Infection at the optimal temperatures of 15–20° C occurs in the presence of free water or at least 90% relative humidity after approximately 15 hr. More time is required at lower temperatures.

Fig. 13. Conidia on conidiophore of *Botrytis cinerea.* (Courtesy R. Wind)

Hyphae generally penetrate directly through the epidermis of susceptible organs. However, wounds facilitate infection, particularly injuries caused by insects, powdery mildew, hail, or birds. Scanning electron microscopy has shown that conidial germ tubes penetrate berries through numerous microfissures that form around nonfunctional stomata.

Under certain conditions, the ovary is infected through the stigma and style at the end of bloom, but the infection remains latent until *véraison.*

Control

Cultivars differ in susceptibility to Botrytis bunch rot based on the compactness of their clusters, the thickness and anatomy of the berry skin, and their chemical composition (anthocyanins and phenolic compounds). It is also known that the vine synthesizes phytoalexins (resveratrol and the viniferins) and that the concentration of these protective substances is related to the relative resistance of cultivars.

Susceptible cultivars usually need to be protected against bunch rot by a combination of cultural practices and chemical control. To slow the development of the disease, avoid excessive vegetation through rootstock management and the judicious use of nitrogen fertilization; increase aeration and exposure of clusters to the sun by using appropriate trellising systems and by removing leaves around the fruit; and provide protection against diseases and insect pests capable of injuring the berries, particularly grape berry moths.

Chemical control is usually necessary but can be conducted only with preventive treatments. A program of four applications (known as the "standard" method in Europe) has given satisfactory results: first treatment at the end of bloom and the beginning of fruit set; second treatment just before berry touch; third treatment at the beginning of *véraison;* and a fourth treatment three weeks before harvest. Chemical treatment may be ineffective if strains of *B. cinerea* develop resistance to the fungicide, as has happened with the benzimidazoles and dicarboximides. Proper adjustment of spraying equipment to give consistently good penetration and coverage of clusters is essential.

Bunch rot in stored table grapes is generally controlled by sulfur dioxide fumigation combined with storage at low temperatures (near 0° C).

With a recently developed mathematical model of the epidemiological behavior of *B. cinerea* on the vine, the risk of disease can be predicted at any given moment, and hence the appropriateness of a chemical treatment can be evaluated. Other research indicates that an antagonistic fungus, *Trichoderma harzianum,* may be an effective biological means of controlling *B. cinerea.* A strategy of integrated control using both the antagonist and chemical sprays appears promising.

Selected References

Bulit, J., and Dubos, B. 1982. Epidémiologie de la pourriture grise. Bull. OEPP/EPPO Bull. 12:37–48.

Bulit, J., and Lafon, R. 1977. Observations sur la contamination des raisins par le *Botrytis cinerea* Pers. Pages 61–69 in: Travaux Dédiés à G. Viennot-Bourgin. Société Française de Phytopathologie, Paris. 416 pp.

Coley-Smith, J. R., Verhoeff, K., and Jarvis, W. R. 1980. The Biology of *Botrytis.* Academic Press, New York. 318 pp.

Dubos, B., Jailloux, F., and Bulit, J. 1982. L'antagonisme microbien dans la lutte contre la pourriture grise de la vigne. Bull. OEPP/EPPO Bull. 12:171–175.

Hill, G., Stellwaag-Kittler, F., Huth, G., and Schlösser, E. 1981. Resistance of grapes in different developmental stages to *Botrytis cinerea.* Phytopathol. Z. 102:328–338.

Jarvis, W. R. 1977. *Botryotinia* and *Botrytis* Species: Taxonomy, Physiology and Pathogenicity. Monogr. 15. Canada Department of Agriculture, Ottawa, Ontario. 195 pp.

McClellan, W. D., and Hewitt, W. B. 1973. Early Botrytis rot of grapes:

Time of infection and latency of *Botrytis cinerea* Pers. in *Vitis vinifera* L. Phytopathology 63:1151–1157.
Pezet, R., and Pont, V. 1986. Infection florale et latence de *Botrytis cinerea* dans les grappes de *Vitis vinifera* (var. Gamay). Rev. Suisse Vitic. Arboric. Hortic. 18:317–322.
Strizyk, S. 1983. Modélisation. La gestion des modèles "EPI." Phytoma 350:13–19.

(Prepared by J. Bulit and B. Dubos)

Black Rot

Black rot is one of the most economically important diseases of grape in the northeastern United States, Canada, and parts of Europe and South America. The disease is indigenous to North America and was probably introduced to other countries via contaminated propagation material. It was introduced into France on phylloxera-resistant stock. Black rot was seen in 1804 in a Kentucky vineyard; however, the first detailed account of the disease was given by Viala and Ravaz in 1886. Crop losses can range from 5 to 80%, depending on the severity of the epidemic, which is governed by inoculum level, weather, and cultivar susceptibility.

Symptoms

All new growth is susceptible to attack during the growing season. Young leaf laminae, petioles, shoots, tendrils, and peduncles can be infected. The main symptom on leaves is the appearance of small, tan, circular spots on the lamina in spring and early summer. Leaf spots appear one to two weeks after infection. Spots vary from 2 to 10 mm in diameter (Plate 21). Lesions become cream colored, with the color deepening to tan and ultimately to reddish brown on the adaxial surface. Leaf spots are bordered by a narrow band of dark brown tissue. Pycnidia develop in the center of these necrotic spots and appear as small, blackish pimples (Plate 22).

Lesions develop on petioles at about the same time leaf lesions appear. Some lesions enlarge, girdle the petiole, and kill the entire leaf. Lesions on peduncles and pedicels are small, darkened depressions, which soon turn black.

Elongated black cankers develop on young shoots throughout the season. Lesions vary in length from a few millimeters to 2 cm. Pycnidia are commonly observed on these lesions. Numerous cankers result in blighting of the growing tips of shoots.

The first indication of infection on berries is the appearance of a small (about 1 mm in diameter) whitish dot. In a matter of hours, this dot is surrounded by a reddish brown ring, which can grow to over 1 cm in diameter within one day. Within a few days, the berry begins to dry, shrivel, and wrinkle until it becomes a hard, blue-black mummy (Plates 23 and 24). The entire cluster may be affected.

On berries of muscadine grapes (*V. rotundifolia*), symptoms appear as small, black, superficial, scabby lesions 1–2 mm in diameter (Plate 25). These lesions do not spread or cause decay of maturing berries as on bunch grapes, although infected young berries may drop or mummify. Lesions may coalesce to form a brown to black crust covering a large part of the surface of a berry. The skin of the infected berry often splits at the edge of larger lesions. The surface of the lesion is cracked and roughened with embedded pycnidia.

Causal Organism

Guignardia bidwellii (Ellis) Viala & Ravaz (anamorph *Phyllosticta ampelicida* (Engleman) Van der Aa), the cause of black rot, produces ascocarps (pseudothecia) in a stroma on overwintered mummies. Pseudothecia (Fig. 14) are separate, black, and spherical (61–199 μm in diameter), with a flat or papillate ostiole at the apex. The centrum is pseudoparenchymatous and without paraphyses.

Asci (36–56 × 12–17 μm) are fasciculate, cylindrical to clavate, short-stipitate, and eight-spored (Fig. 14). The ascus wall is thick and composed of two layers. Ascospores (10.6–18.4 × 4.8–9.0 μm) are hyaline, nonseptate, oval or oblong, straight or inequilateral, rounded at the ends, and biseriate and are often surrounded by a mucilaginous sheath.

Black, spherical pycnidia (Fig. 14) 59–196 μm in diameter are produced on the host during the growing season. They are solitary, erumpent, and ostiolate at the apex. On leaf blades they occur in circular, reddish brown, necrotic spots. On stems, tendrils, peduncles, and petioles, they are found in elliptical to elongate, brown to black cankers. On fruits they are present in berry mummies or in brown to black, superficial scabs and cankers.

Conidia are hyaline, nonseptate, ovoid to oblong, and rounded at the ends; they measure 7.1–14.6 × 5.3–9.3 μm. Spermagonia are black, spherical (45–78 μm in diameter), innate, erumpent, and ostiolate at the apex. They are produced toward the end of the growing season on berry mummies and dead leaves in association with ascogonial stromata. Spermatia are hyaline, nonseptate, and bacilliform and measure 2.5 × 1 μm.

A distinct physiologic race of the fungus, differing in pathogenicity from *G. bidwellii* on American bunch grapes, occurs on muscadine grapes. *G. bidwellii* "f. *euvitis* Luttrell" is pathogenic to American species of the *Vitis* section *Euvitis* and to *V. vinifera*. *G. bidwellii* f. *muscadinii* Luttrell is pathogenic to *V. rotundifolia* and *V. vinifera*. A third race, *G. bidwellii* "f. *parthenocissi* Luttrell," is pathogenic only to *Parthenocissus* spp. Besides differing in pathogenicity, the race on muscadine grapes also differs in appearance, growth rate of colonies in culture, and size of pseudothecia, ascospores, and conidia.

Disease Cycle and Epidemiology

The fungus overwinters in mummified berries on the soil or in

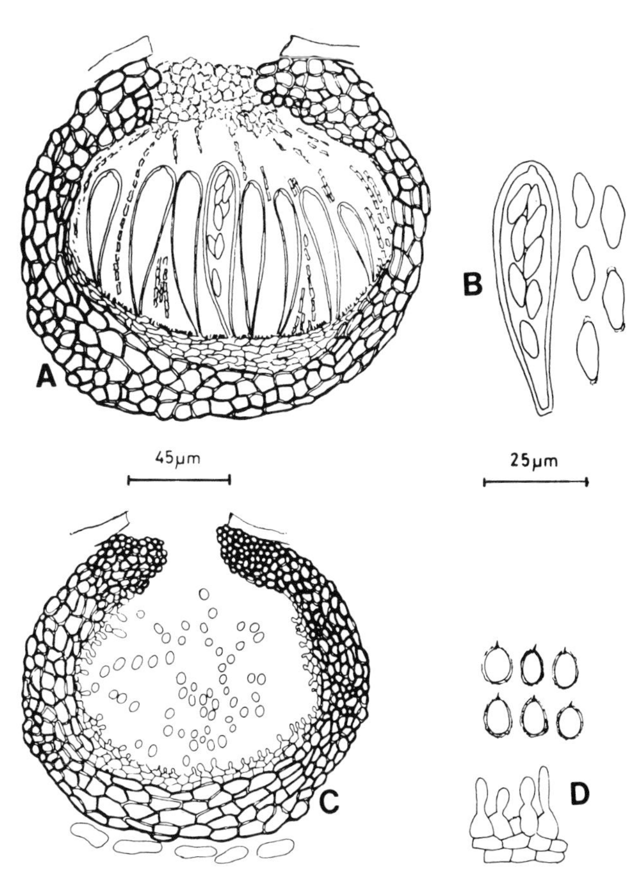

Fig. 14. Fruiting bodies and spores of *Guignardia bidwellii*. **A,** Cross section through a pseudothecium, showing asci within a locule. **B,** Ascus and ascospores. **C,** Cross section through a pycnidium. **D,** Conidiogenous cells and conidia. (Modified and reprinted, by permission, from Sivanesan and Holliday, 1981)

old clusters still hanging in the vines (Fig. 15). Ascospore discharge begins in spring shortly after budbreak. Ascospores are ejected after 0.3 mm or more of rain and may be discharged for up to 8 hr after one rainfall. Discharge continues during rains through mid-July but then diminishes.

Ascospores cause leaf lesions and also infect blossoms and young fruit. Fruit infections occur from midbloom until onset of berry color. Mature leaves and ripe fruit are not susceptible.

Ascospores require free water for germination. They germinate in as little as 6 hr at 27° C, which is also optimal for leaf infection. Longer wet periods are required for infection at 10–21° C. No infection occurs at 32° C.

Pycnidia develop in overwintered mummies and newly rotted berries and can develop in leaf lesions three to five days after infection. Once pycnidia mature, conidia are liberated following 3 mm or more of rain. Vast numbers of conidia are liberated from pycnidia in leaf lesions and rotted fruit during the growing season and cause secondary infection. Rain lasting for 1–3 hr is optimal for dispersal of conidia. Environmental conditions required for conidial germination and infection are similar to those for ascospores. Conidia can attack leaves, blossoms, and young fruit. Peak fruit infection occurs at midbloom on Concord grapes in Michigan. Very few fruit or leaves are infected after late July, and none are infected after the end of August. Leaf infections occur after 6 hr of wetness at 26.5° C but require 24 and 12 hr of wetness at 10 and 32° C, respectively.

The black rot fungus on muscadine grapes survives the winter in the perithecial state in overwintering leaves and in the pycnidial state in infected stems. Spermagonia and perithecial initials develop in dead overwintering leaves from October to December. Asci mature in the perithecia in late winter and early spring, and ascospores are discharged for four to five weeks in April and May. The ascospores and conidia are the primary inoculum and are spread to new growth by air currents and blowing rain, respectively. Immature leaves are infected during wet periods throughout the growing season. Immature berries may be infected from the time they are set until they reach full size. Conidia are released from pycnidia during the growing season and initiate secondary infections.

Control

Chemical control of black rot has been based on the use of protective fungicides, such as maneb or ferbam, starting when shoots are 10–16 cm long and continuing until the berries contain about 5% sugar. In areas where disease is generally severe, earlier sprays may be necessary. Curative fungicides such as triadimefon applied after infection have shown promise.

Black rot of muscadine grapes can be effectively controlled by applying protective fungicides such as maneb or captan, beginning after bloom and repeating at 14-day intervals until August.

Removing overwintering mummified berries from the vine and disking mummies into the soil are beneficial control practices in *Euvitis.*

Species of *Vitis* and cultivars within species differ in their susceptibility to black rot. Some species of *Vitis,* in descending order of susceptibility to black rot, are as follows: *V. vinifera* (very susceptible), *V. arizonica, V. californica, V. labrusca, V. rubra, V. monticola, V. coriacea, V. aestivalis, V. rupestris* (St. George), *V. berlandieri, V. cordifolia, V. riparia,* and *V. candicans* (very resistant).

Selected References

Clayton, C. N. 1975. Diseases of muscadine and bunch grapes in North Carolina and their control. N.C. Agric. Exp. Stn. Bull. 451. 37 pp.

Ferrin, D. M., and Ramsdell, D. C. 1977. Ascospore dispersal and infection of grapes by *Guignardia bidwellii,* the causal agent of grape black rot disease. Phytopathology 67:1501–1505.

Ferrin, D. M., and Ramsdell, D. C. 1978. Influence of conidia dispersal and environment on infection of grape by *Guignardia bidwellii.*

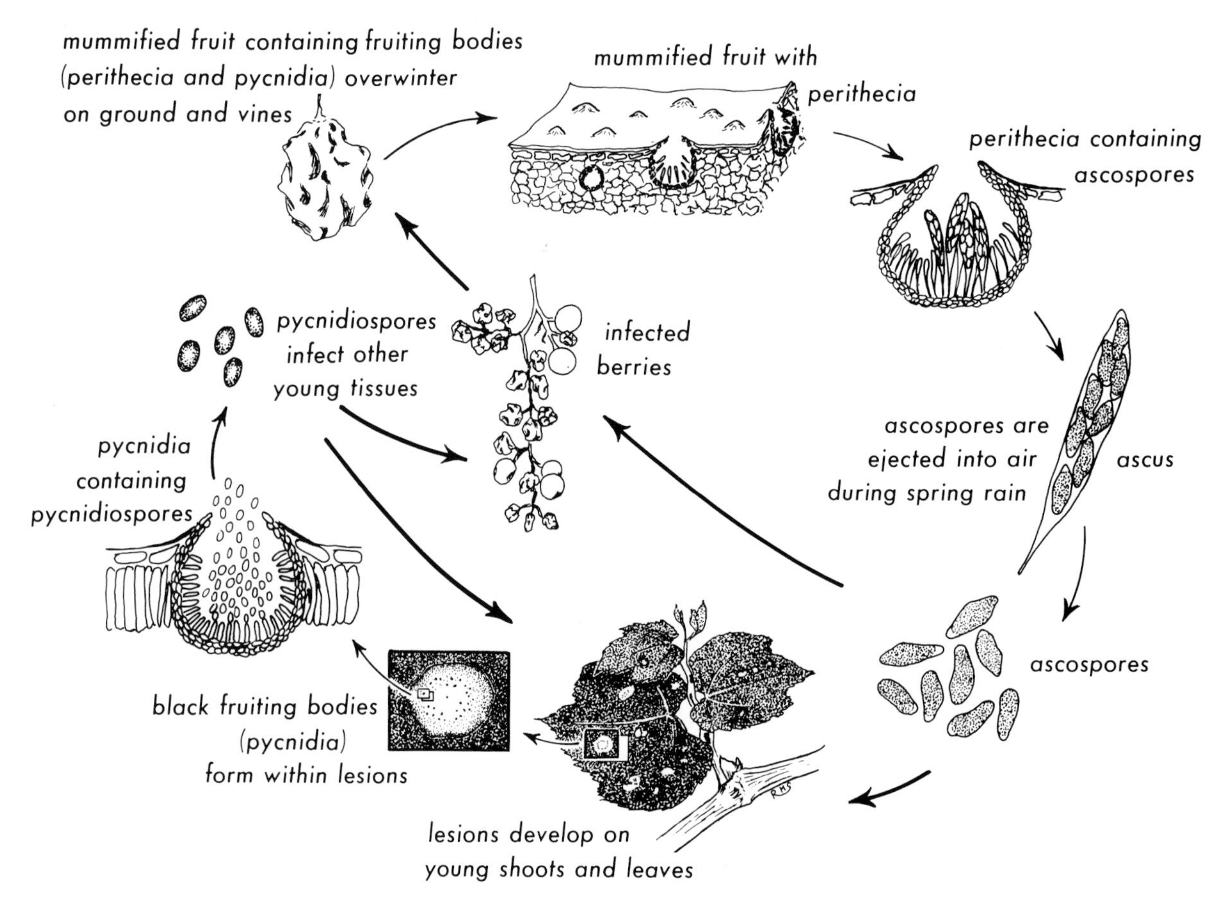

Fig. 15. Disease cycle of black rot. (Drawing by R. Sticht)

Phytopathology 68:892–895.
Luttrell, E. S. 1946. Black rot of muscadine grapes. Phytopathology 36:905–924.
Luttrell, E. S. 1948. Physiologic specialization in *Guignardia bidwellii*, cause of black rot of *Vitis* and *Parthenocissus* species. Phytopathology 38:716–723.
Sivanesan, A., and Holliday, P. 1981. *Guignardia bidwellii*. Descriptions of Pathogenic Fungi and Bacteria, No. 710. Commonwealth Mycological Institute, Kew, Surrey, England.
Spotts, R. A. 1977. Effect of leaf wetness duration and temperature on the infectivity of *Guignardia bidwellii* on grape leaves. Phytopathology 67:1378–1381.
Spotts, R. A. 1980. Infection of grape by *Guignardia bidwellii*—Factors affecting lesion development, conidial dispersal, and conidial populations on leaves. Phytopathology 70:252–255.

(Prepared by D. C. Ramsdell and R. D. Milholland)

Phomopsis Cane and Leaf Spot

Phomopsis cane and leaf spot, referred to as *excoriose* in Europe and as "dead arm" in American literature before the description of Eutypa dieback, is widely distributed throughout the viticultural world. It has been reported in Africa, Asia, Australia and Oceania, Europe, and North America.

The disease is especially destructive in regions where the climate following budbreak may keep grapevines wet by rain for several days. It can weaken vines, reduce yield, lower the quality of fruit for table use, and kill grafted and other nursery stock.

Symptoms

Infected leaf blades have small, light green or chlorotic, irregular to circular spots with dark centers. They may be puckered along veins near the perimeter, or the margin may be turned under. Dark brown to black necrotic spots may also occur along primary and secondary leaf veins and on petioles. The necrotic spots may drop out of the leaf, causing a "shot-hole" appearance. Infected portions of the leaf may turn yellow, then brown (Plate 26). Severely infected leaves and leaves with heavily infected petioles usually abscise.

Infected young shoots, cluster stems (rachises), and petioles have chlorotic spots with dark centers. As these spots enlarge, the infected tissues turn dark brown to black and appear as streaks and blotches. When infections on a shoot are numerous, they often coalesce to form dark blotches, which may involve much of the surface of the basal three to six internodes (Plate 27). During rapid growth of shoots, these dark, necrotic blotches often crack and become open fissures in the cortex tissue. Cracks in the epidermis and cortex of shoots tend to heal during the growing season and become rough as the tissues mature. Cluster stems may blight and become brittle from numerous infections, resulting in breakage of the cluster and loss of fruit.

By midseason, symptoms become obscure because of vine growth and leaf cover. Although symptoms are generally seen on portions of the basal three to six internodes, they also can be found on two or more internodes at intervals along some shoots. These infections occurred at the shoot tip during successive rain-induced infection periods.

The fungus also causes a fruit rot (Plate 28). Infections appear to be associated with lenticels. It has also been suggested that mycelium advances into the berry from a lesion on the pedicel (cap stem). Infected fruit gradually turn brown and shrivel. Pycnidia develop sparsely in the epidermis of the berry. Some grapes of highly susceptible cultivars of *V. vinifera* (Kandahar, Olivette blanche, Olivette noir, Rish Baba, and Flame Tokay) may become infected through the skin when very young and show blackish flecks in the skin. When the grapes are mature, the fungus in some of the black-fleck lesions resumes growth and rots the fruit. Most fruit infections, however, come from lesions on the rachis or pedicel. Infected fruits may abscise from the pedicel, leaving a dry stem scar.

In winter, pycnidia and irregular dark blotches (2 × 3 cm) with irregular light centers occur on the surface of infected canes. Pycnidia become prominent in the cortex of diseased one-year-old canes (Plate 27), spurs, rachis stubs, old tendrils, and petioles. Pycnidia may be so numerous that they lift the epidermis and admit air under it, which gives the surface a white to silvery sheen.

Causal Organism

Phomopsis viticola (Sacc.) Sacc. (syn. *Fusicoccum viticola* Reddick) is the causal agent of Phomopsis cane and leaf spot. An ascogenous stage, *Cryptosporella viticola* Shear, has been described but appears to be rare, and its role in the epidemiology of the disease is uncertain.

P. viticola produces black pycnidia ranging in diameter from 0.2 to 0.4 mm. Each pycnidium has one or more locules, depending on the substrate. Pycnidia are discoid in early stages of development and become globose at maturity. An ostiole opens at the peak of a short neck. The opening on a pycnidium is usually round and smooth but may be irregular and sometimes ragged.

Pycnidiospores arise through the ostiole, either in long, curled, yellowish to cream-colored cirri (Plate 29) or in a gelatinous mass. The inner surface of a pycnidium is lined with straight pycnidiophores of two types. One type has a pointed apex, measures 12–20 × 2 μm, and bears single, hyaline, elliptical-fusoid pycnidiospores or alpha spores (7–10 × 2–4 μm) that are acute on one or both ends and normally have a large guttule at each end (Fig. 16). The other type of pycnidiophore is short (5–8 × 1.5 μm) and produces scolecospores or beta spores, which are long, curved, threadlike or wormlike spores measuring 18–30 × 0.5–1 μm (Fig. 16). The function of scolecospores, which are not known to germinate, has not been determined.

In green shoots, the mycelium invades mostly cortex

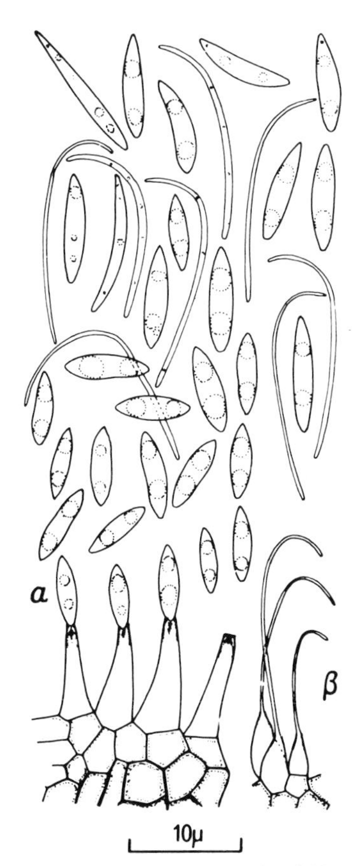

Fig. 16. Conidiogenous cells and conidia of *Phomopsis viticola*, showing alpha (α) and beta (β) spores. (Reprinted, by permission, from Punithalingam, 1979)

parenchyma tissue. The mycelium is clear and forms pseudoparenchymatous mats among host cells. The mats may turn black, making spots and lesions appear black. Pycnidia form in the black areas about 14 days after infection.

The presence of *P. viticola* can easily be determined by placing a piece of diseased tissue with mature pycnidia in a moist chamber. Mature pycnidia will exude spores in cirri or gelatinous masses within 24–48 hr (Plate 29).

In culture, mycelium of *P. viticola* is hyaline, septate, and branched and forms a dense mat, often in concentric rings under alternating light-dark regimes. Portions of the mat or the margins of concentric rings turn black as the culture ages. The mats are often sectored in white and black mycelium. Pycnidia form irregularly in the dark parts of the mycelial mat, either singly or in clumps.

The perithecia of *C. viticola* are buried in irregularly pulvinate subcortical stromata. They are thin-walled and globose, with a short, stout, smooth beak. Asci (60–72 × 7–8 μm) are sessile or subsessile. Slender, septate, wavy paraphyses protrude beyond the asci. Ascospores (11–15 × 4–6 μm) are subelliptic, obtuse, hyaline, and single-celled.

Disease Cycle and Epidemiology

P. viticola overwinters as mycelium and pycnidia in bark. It has also been reported to occur as mycelium in dormant buds. In spring, mature pycnidia erupt through the epidermis of canes, leaf petioles, and other dead, diseased parts hanging in the vine, as well as through cracks in bark on older diseased tissue. When wet, spores exude from the pycnidia and are washed or splashed by rain to shoot tips.

Alpha spores germinate in a temperature range of 1–37° C. At the optimal temperature of 23° C, infection may take place within a few hours in free water or near 100% relative humidity. Only very young tissues are infected. Symptoms appear 21–30 days after infection. In summer in warm, dry climates, the fungus usually becomes inactive, but in fall as the season cools, it resumes activity. In cool climates the fungus may remain active throughout the growing season.

Where the disease is endemic, it may become severe when rain or showers continue for several days during early spring. When the mean temperature is 5–7° C, shoot growth slows, and shoots that are 3–10 cm long are very susceptible to infection. Prolonged periods of rain and cold weather are prime factors in the development of an epidemic. Because of the buildup of inoculum, the disease becomes increasingly severe with each successive cool, wet spring.

Because the fungus spreads mostly within the vine rather than from vine to vine, spread within the vineyard is localized, remaining in close proximity to the inoculum source. Long-distance spread occurs by transport of infected or contaminated propagation materials such as budwood, cane cuttings, and nursery stock.

Control

Phomopsis cane and leaf spot can be controlled by a combination of sanitation and fungicide application. To avoid introducing *P. viticola* into the vineyard, use pathogen-free propagation materials (cuttings, buds, rootings, and grafted or budded nursery stock) when planting or replanting. Once the disease has appeared, remove as much diseased and dead wood as practical during pruning. Destroy debris after pruning by shredding it, disking or plowing it into the soil, or burning it.

Where permitted, eradicant chemical sprays, such as sodium arsenite or dinoseb, may be applied during late dormancy (two to three weeks before bud swell) to kill the pycnidia and spores on the surface of vine parts. Where permitted, 8-hydroxyquinoline sulfate is used to disinfect propagation material.

Two applications of a protective chemical are generally recommended, the first when the shoots are 1–3 cm long and the second when the shoots average 6–12 cm long. Captan, folpet, and maneb are effective chemicals. When the temperature is cool and shoot growth is slow, one or more additional applications may be necessary.

Few if any cultivars of grapes are resistant to Phomopsis cane and leaf spot. Nevertheless, cultivars differ widely in relative susceptibility, which also seems to vary with locality.

Selected References

Bugaret, Y. 1986. Données nouvelles sur l'épidémiologie de l'excoriose et leurs conséquences pour la lutte. Phytoma 375:36–41.

Bulit, J., Bugaret, Y., and Lafon, R. 1972. L'excoriose de la vigne et ses traitements. Rev. Zool. Agric. Patol. Veg. 1:44–54.

Doazan, J. P. 1974. Sensibilité de variétés de la vigne (*V. vinifera* L.) à l'excoriose (*Phomopsis viticola* Sacc.). Distribution du caractère dans quelques descendances. Vitis 13:206–211.

Gärtel, W. 1972. *Phomopsis viticola* Sacc., der Erreger der Schwarzfleckenkrankheit der Rebe (dead-arm disease, Excoriose) - seine Epidemiologie und Bekämpfung. Weinberg Keller 19:13–79.

Gregory, C. T. 1913. A rot of grapes caused by *Cryptosporella viticola.* Phytopathology 3:20–23.

Pezet, R. 1976. L'excoriose de la vigne: Généralités et connaissances nouvelles. Rev. Suisse Vitic. Arboric. Hortic. 8:19–26.

Pine, T. S. 1958. Etiology of the dead-arm disease of grapevines. Phytopathology 48:192–197.

Pine, T. S. 1959. Development of the grape dead-arm disease. Phytopathology 49:738–743.

Punithalingam, E. 1979. *Phomopsis viticola.* Descriptions of Pathogenic Fungi and Bacteria, No. 635. Commonwealth Mycological Institute, Kew, Surrey, England.

(Prepared by W. B. Hewitt and R. C. Pearson)

Anthracnose

Anthracnose or bird's-eye rot is a disease of European origin. Before the introduction of powdery mildew and downy mildew, anthracnose was the most damaging grape disease in Europe. Anthracnose has been reported from all grape-growing countries, and it was probably carried from Europe on propagation material. Anthracnose is a disease of rainy, humid regions, where some grape cultivars are practically impossible to grow because of the disease. Because of this environmental requirement, anthracnose does not occur on the West Coast of the United States, although it can be a problem east of the Rocky Mountains.

After the introduction of Bordeaux mixture in 1885, the incidence and severity of anthracnose on European cultivars of *V. vinifera* were significantly reduced, but the disease was still observed on some interspecific hybrids and rootstocks that were not sprayed routinely. More recently, as copper compounds have been replaced by copper-free organic fungicides, renewed outbreaks of anthracnose have been observed in many regions.

Anthracnose reduces the quality and quantity of the crop and weakens the vine.

Symptoms

Anthracnose produces circular (1–5 mm in diameter) leaf lesions, with brown to black margins and round or angular edges. The lesions are often quite numerous and may coalesce or remain isolated (Plate 30). The center of the lesions becomes grayish white and dries. The necrotic tissue eventually drops out of the lesion, leaving a "shot-hole" appearance. Young leaves are most susceptible to infection. Lesions may cover the entire blade or appear mainly along the veins. When the veins are affected, especially on young leaves, the lesions prevent normal development, resulting in malformation or complete drying of the leaf. Because the youngest leaves are the most susceptible, the malformations are most obvious at the tips of the shoots, which appear burned.

Young, green, succulent parts of the shoot are most susceptible to anthracnose. Lesions on shoots are small and isolated, with round or angular edges (Plate 31). Lesions have a violet-brown margin, which gradually becomes violet-black. Lesions may coalesce. The center of the lesions may extend to the pith of the shoot. Callus forms around the edge of the lesions. These lesions on the shoots may crack, causing the shoots to become brittle. Anthracnose lesions on the shoots may be confused with hail injury; however, unlike hail damage, the edges of the wounds caused by the anthracnose fungus are raised and black. Anthracnose on petioles appears similar to that on the shoots.

Clusters are susceptible to infection before flowering and until *véraison*. Lesions on the rachis and pedicels appear similar to those on shoots. If the rachis is girdled, the distal portion of the cluster may shrivel. Lesions on berries are surrounded by a narrow, dark brown to black margin (Plate 32). The center of the lesion is violet in the early stages but gradually becomes velvety and whitish gray. Lesions on berries may extend into the pulp, which induces cracking.

Causal Organism

The causal organism of anthracnose is *Elsinoë ampelina* (de Bary) Shear (syn. *E. viticola* Raciborski, anamorph *Sphaceloma ampelinum* de Bary [syns. *Gloeosporium ampelophagum* (Pass.) Sacc., *Ramularia ampelophaga* Pass.]). The fungus produces acervuli on the exterior of lesions. Acervuli contain short, cylindrical, and crowded conidiophores bearing small, ovoid, hyaline conidia (3–6 × 2–8 μm) with mucilaginous walls and one or two refringent spots. In water, conidia produce germ tubes that readily fasten to the substrate. In autumn, production of acervuli ceases and sclerotia form at the edge of lesions on shoots. These sclerotia are the main overwintering survival structures. In spring, conidia are produced on the sclerotia.

Asci (80–100 × 11–23 μm) are formed in pyriform locules of ascostromata. Asci contain eight brown-black, four-celled ascospores (29–35 × 4.5–7 μm). The ascospores germinate at 2–32° C, infect tissue, and develop lesions, which give rise to the *Sphaceloma* stage.

Disease Cycle and Epidemiology

In spring, overwintered sclerotia produce numerous conidia when wet for 24 hr or more at temperatures above 2° C. Thereafter, 2 mm or more of rain disseminates conidia to green tissue, where they germinate to cause primary infection when free water is present for at least 12 hr. Conidia can germinate and infect at 2–32° C; the subsequent incubation period varies from 13 days at 2° C to four days at 32° C. The optimal temperatures for disease development are 24–26° C.

Conidia or ascospores formed on infected berries overwintered on the vineyard floor may also cause primary infections.

Temperature and moisture are the main environmental factors influencing disease development. Anthracnose is especially damaging during years of heavy rainfall. In the summer, hail may favor the spread of the disease.

Control

Planting highly susceptible cultivars such as Afuz Aly, Thompson Seedless (Sultanina), Regina Nera, Cardinal, Delicia, Citronella, Black Corinth, Italia, Pedro Ximenes, and Queen of the Vineyards is not recommended in heavy soils with poor drainage. Phytosanitary regulations that prohibit transport of infected propagation material should be observed.

Dormant-season sprays of DNOC, DNBP, Bordeaux mixture, or lime sulfur are recommended in some viticultural regions to control anthracnose. During the growing season, foliar applications of fungicide are recommended at two-week intervals beginning when shoots are 5–10 cm long. Sprays are also recommended within 24 hr after hail or sprinkler irrigation.

Selected References

Arnaud, G., and Arnaud, M. 1931. Traité de Pathologie Végétale. 2 vols. Lechevalier et Fils, Paris. 1,831 pp.

Brook, P. J. 1973. Epidemiology of grapevine anthracnose, caused by *Elsinoë ampelina*. N.Z. J. Agric. Res. 16:332–342.

Carne, W. M. 1926. Black rot or anthracnose of the grape vine (*Gloeosporium ampelophagum*). J. Dep. Agric. West. Aust. Ser. III 2:178–182.

du Plessis, S. J. 1940. Anthracnose of vines and its control in South Africa. Sci. Bull. 216. Department of Agriculture, Pretoria, South Africa.

Miricǎ, I., and Miricǎ, A. 1981. Antracnoza viţei de vie şi combaterea ei, studiu monografic. Editura Ceres, Bucharest. 162 pp. (In Rumanian, English summary)

Shear, C. L. 1929. The life history of *Sphaceloma ampelinum* de Bary. Phytopathology 19:673–679.

Sivanesan, A., and Critchett, C. 1974. *Elsinoë ampelina*. Descriptions of Pathogenic Fungi and Bacteria, No. 439. Commonwealth Mycological Institute, Kew, Surrey, England.

Sutton, B. C., and Pollack, F. G. 1973. *Gloeosporium cercocarpi* and *Sphaceloma cercocarpi*. Mycologia 65:1125–1134.

(Prepared by I. I. Miricǎ)

Rotbrenner

Symptoms of and severe losses from *Rotbrenner* were well known in Europe during the last century. Before Müller-Thurgau first described the causal agent, the disease was thought to be caused by unfavorable weather conditions, water deficiency, or excessive soil moisture. Although *Rotbrenner* has been identified in viticultural regions of most European countries, it is generally confined to certain locally restricted areas. In some areas the disease results in severe losses annually, whereas in others it occurs sporadically or not at all. The disease has also been observed to occur during several succeeding years in certain locations and then not appear for several years.

The fungus attacks the cultivated grape, *V. vinifera*, as well as other species and interspecific hybrids of *Vitis*. The disease may also occur on Virginia creeper (*Parthenocissus quinquefolia*) and Boston ivy (*P. tricuspidata*).

Symptoms

Lesions on leaves are initially yellow on white-fruited cultivars (Plate 33) and bright red to reddish brown on red- and black-fruited (*V. vinifera*) cultivars. A reddish brown necrosis then develops in the center of the lesion, leaving only a thin margin of yellow or red tissue between the necrotic and green areas of the leaf. The lesions are typically confined by the major veins and the edge of the leaf and may be several centimeters wide. Atypical symptoms—freckled spots or discoloration scattered over the leaf surface—often occur late in the season.

Early infections may occur on the first to the sixth leaves of young shoots, resulting in minor losses. Later infections may attack leaves up to the 10th or 12th position on the shoot and can result in severe defoliation.

In addition, the fungus may attack inflorescences before or during bloom, causing them to rot and dry out (Plate 34). Unlike in Botrytis blight, only the pedicels are attacked; the fungus does not seem to invade the rachis of the cluster. In severe cases, the berries are destroyed, leaving the rachis of the cluster bearing only a few individual berries or none at all. High levels of infection during bloom can result in 80–90% crop loss.

Causal Organism

Pseudopezicula tracheiphila (Müll.-Thurg.) Korf & Zhuang (syn. *Pseudopeziza tracheiphila* Müll.-Thurg., anamorph *Phialophora tracheiphila* (Sacc. & Sacc.) Korf [syn. *Botrytis tracheiphila* Sacc. & Sacc.]), the cause of *Rotbrenner*, is a discomycete with minute (up to 0.6 mm in diameter), sessile, gelatinous, whitish to faintly colored, gregarious apothecia. The apothecia are erumpent from leaf tissue and are often associated with the veins.

The inoperculate, broadly clavate, eight-spored asci (115–145 × 18–28 μm) (Fig. 17) show a blue pore in iodine only after pretreatment in aqueous potassium hydroxide. The ellipsoid, hyaline ascospores (19–26 × 9–14 μm) are flattened on one side and produce a sporelike vesicle upon germination, which gives rise to one or more germ tubes or conidiophores. Paraphyses are branched and curved or slightly deformed at the apex, filiform, septate, and hyaline.

On malt agar, the anamorph may be formed, with hyaline, septate, short conidiophores that are coarser than the vegetative hyphae. Conidiogenous cells are monophialidic and lageniform, with well-defined but thin-walled collarettes. Conidia (2–3 × 1.5–2 μm) are ellipsoid, hyaline, and unicellular. The hyphae grow in a characteristic sine-wave pattern (Fig. 18) that, when observed in the vessel elements of diseased tissue, is considered diagnostic by many plant pathologists.

A disease very similar to *Rotbrenner,* called angular leaf scorch, has been described in New York State. The fungus causing angular leaf scorch in North America produces smaller apothecia (0.1–0.3 mm in diameter) than *P. tracheiphila,* and its broadly clavate asci (80–100 × 20–22 μm) have four spores, in contrast to the eight-spored European fungus. The American counterpart has been described as a distinct species, *P. tetraspora* Korf, Pearson & Zhuang (anamorph *Phialophora*-type).

Disease Cycle and Epidemiology

Apothecia are formed primarily on fallen leaves in the spring, although they may develop on current-season infected leaves in late summer or fall. Depending on the weather conditions, ascospores (primary inoculum) may be present throughout the year.

Little is known regarding temperature and moisture conditions necessary for ascospore maturation, release, and germination or for infection. Heavy rainfall and prolonged wetting periods favor infection and lead to severe disease. Young leaves are susceptible after they reach a width of about 5 cm. After an incubation period of two to four weeks, the fungus may invade the vascular elements of infected leaves, causing symptom development. The fungus remains latent if it is unable to invade the vessel elements, in which case it can be isolated from green leaves showing no symptoms. Conditions required for the fungus to invade the vascular system are not well understood; however, soil conditions and water supply that place the vine under temporary stress appear to be important factors.

Control

Effective fungicides should be applied when the first leaf reaches a width of about 5 cm; applications should be repeated at intervals of seven to 10 days, depending on plant development. It is especially important to prevent the disease during bloom. Later in the season, the control of *Rotbrenner* can be combined with sprays against downy mildew. Dithiocarbamates are the most effective fungicides for control of *Rotbrenner.*

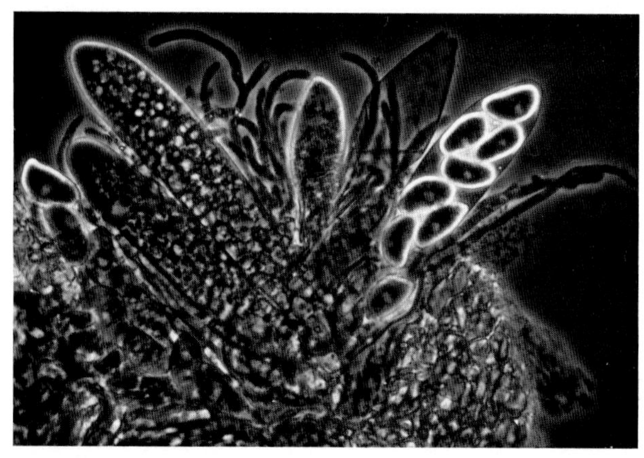

Fig. 17. Eight-spored asci and branched paraphyses of *Pseudopezicula tracheiphila.* (Courtesy W. Gärtel)

Selected References

Korf, R. P., Pearson, R. C., Zhuang, W., and Dubos, B. 1986. *Pseudopezicula* (Helotiales, Peziculoideae), a new discomycete genus for pathogens causing an angular leaf scorch disease of grapes ("Rotbrenner"). Mycotaxon 26:457–471.

Levadoux, M. L. 1944. Le brenner (*Pseudopeziza tracheiphila* Müll.-Thurg.). Bull. Off. Int. Vin 17:43–54.

Müller-Thurgau, H. 1903. Der rote Brenner des Weinstockes. Centralbl. Bakteriol. Parasitenkd. Infektionskr. Abt. 2 10:1–38.

Schüepp, H. 1976. Verstärktes Auftreten des Rotbrenners der Rebe. Schweiz. Z. Obst Weinbau 112:379–381.

Siegfried, W., and Schüepp, H. 1983. Der Rotbrenner der Rebe, eine unberechenbare Krankheit. Schweiz. Z. Obst Weinbau 119:235–239.

(Prepared by H. Schüepp)

Bitter Rot

Bitter rot is a disease of ripe fruit of the *Euvitis* section of *Vitis* caused by a weakly pathogenic fungus that attacks damaged or near-senescent tissues under warm and humid conditions. Bitter rot is common in the eastern United States south of Pennsylvania and west to Texas. On rare occasions it has been found as far north as the Finger Lakes region of New York. It has been reported from Australia, Brazil, Greece, India, Japan, New Zealand, and South Africa. It has not been reported from France or Germany.

The acceptability of the diseased fruit for either table or wine

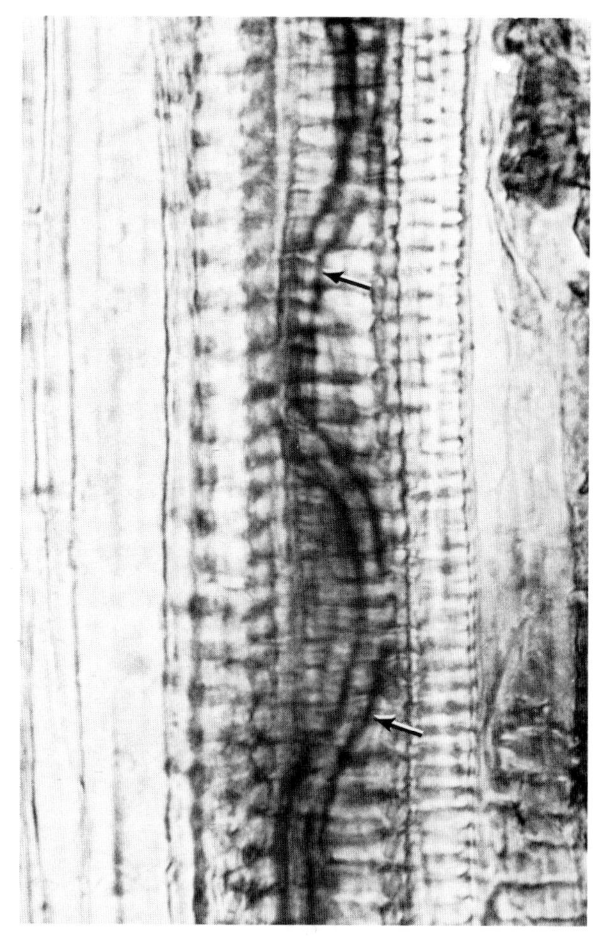

Fig. 18. Hypha (arrows) of *Pseudopezicula tracheiphila* growing in a sine-wave pattern inside a vessel element of a leaf. (Courtesy H. Schüepp)

use is markedly reduced. The bitter flavor is carried through the winemaking process and gives the wine an unpleasant, burnt-bitter taste.

The bitter rot fungus causes a severe disease on muscadine grapes. It attacks all green tissues and can be parasitic throughout the growing season.

Symptoms

The fungus usually invades a berry from the pedicel. Light-colored berries become brownish and often show concentric rings of acervuli (Plate 35) before the entire berry is involved. On blue berries, the surface may take on a roughened, sparkly appearance as the acervuli begin to develop. Within a couple of days, the berry softens and is easily detached. At this stage, the bitter taste of the berry is most pronounced. Berries that do not shell continue to dry, become firmly attached, and are less obviously bitter in taste. When shriveled, the rotted berries look much like berries affected by black rot, ripe rot, or the fruit-infection stage of Phomopsis cane and leaf spot.

The acervuli of the bitter rot fungus are large and irregular in outline (Fig. 19). They form after the berry has reached full size, causing the epidermis and cuticle to rupture. The ruptured epidermis is less obvious after the berry has been wet and secondary organisms have grown over its surface.

The fungus has been reported to cause a girdling of the shoots of several cultivars of *V. vinifera* throughout the growing season in Greece. A minor flecking of young leaves and shoots of Warren (*V. bourquina*) and Concord (*V. labrusca*) grapes caused by the same fungus has been reported from Georgia in the United States.

The disease has many symptoms on muscadine grapes, including flecking of young leaves, stems, and individual flower buds; olive brown lesions with acervuli on young green berries; and blight of pedicels, which causes the young berries to shrivel and generally break off. As berries approach maturity, the disease may spread rapidly, causing a soft rot.

Causal Organism

Greeneria uvicola (Berk. & Curt.) Punithalingam (syn. *Melanconium fuligineum* (Scribner & Viala) Cav.), the cause of bitter rot, forms acervuli (varying in diameter up to 250 μm) that are separate to confluent and subepidermal, with irregular dehiscence (Fig. 20). Conidiophores (30 × 3 μm) are hyaline, septate, and irregularly branched. Conidiogenous cells are enteroblastic, phialidic, discrete or more frequently integrated, determinate, cylindrical, tapered to the apexes, hyaline, and smooth; the collarette and channel are minute. Conidia (7.5–10 × 3–4 μm) are dark, smooth, thin-walled, nonseptate, cylindrical, and fusiform to oval, with truncate base and obtuse apex (Fig. 20). A perfect stage of the organism has not been described.

Disease Cycle and Epidemiology

G. uvicola is widely distributed. It can overwinter as a saprophyte on senescent and fallen leaves and berries, on cold-damaged tips of shoots, and in the necrotic bark layer of year-old canes within its area of climatic adaptation.

The fungus may invade any injured tissue of *Euvitis* cultivars. For example, acervuli have been found in the necrotic tissue of immature green berries immediately surrounding the emergence hole of the grape berry moth and in necrotic areas on leaves damaged by contact herbicides. The fungus, however, generally does not move beyond the damaged areas into healthy green tissues.

The disease cycle on the fruit starts shortly after flowering, when characteristic corky lenticular warts form on the pedicel of each berry. The fungus invades the dead cells of these warts and remains latent until the berry reaches maturity. It then invades the pedicel and moves into the berry, where conidia are produced within four days. At this stage of the cycle, there is sufficient inoculum potential that any injury to healthy berries permits rapid spread of the disease. Bird pecking, insect injury, or cracking of berries due to rain will permit conidial infection of berries.

Infection occurs from 12°C to an optimum of 28–30°C. Mycelial growth is inhibited above 36°C.

Control

Bitter rot is generally controlled by the fungicide sprays applied to control more serious diseases. However, if the late-season or preharvest sprays are omitted, the potential for bitter rot development is greatly increased. Most commonly used fungicides (captan, ferbam, maneb) are effective against *G. uvicola.*

Most cultivars have moderate to good resistance against

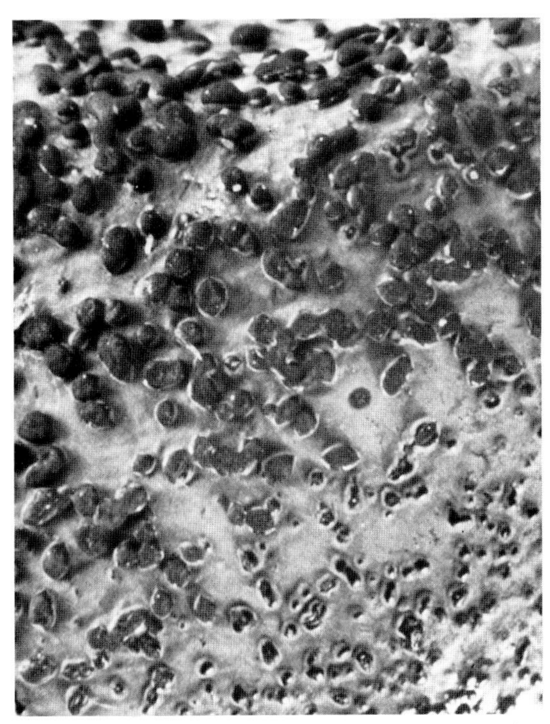

Fig. 19. Acervuli of the bitter rot fungus fruiting on the surface of a berry. (Courtesy J. R. McGrew)

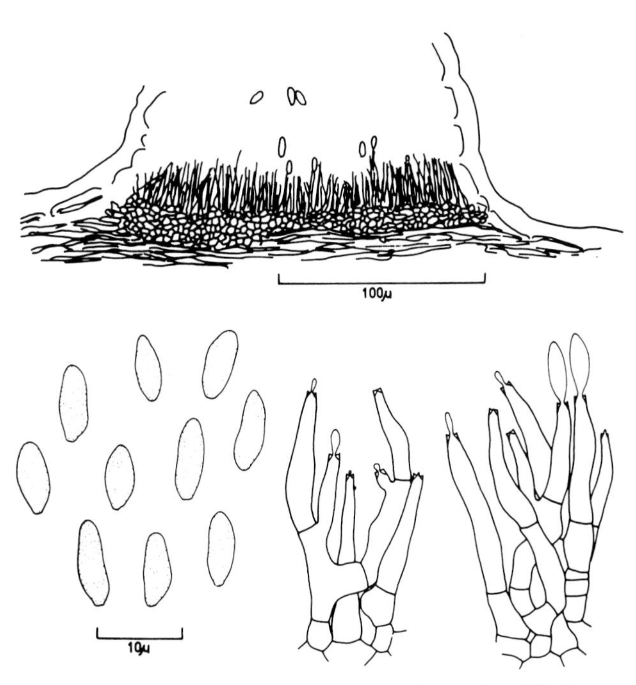

Fig. 20. Acervulus (top), conidia (bottom left), and conidiophores (bottom right and center) of *Greeneria uvicola*. (Reprinted, by permission, from Sutton and Gibson, 1977)

bitter rot. The more susceptible selections and introductions are generally discarded by growers. Susceptibility within any cultivar increases with fruit maturity, so fruit that remains on the vine past normal harvest is vulnerable to infection.

Selected References

Critopoulos, P. D. 1961. Girdling of grapevine canes by *Melanconium fuligineum*. Phytopathology 51:524–528.

Luttrell, E. S. 1953. Melanconium leaf and stem fleck of grapes. Phytopathology 43:347–348.

Reddy, M. S., and Reddy, K. R. C. 1983. Greeneria fruit rot—An endemic disease of grape in India. Indian Phytopathol. 36:110–114.

Ridings, W. H., and Clayton, C. N. 1970. *Melanconium fuligineum* and the bitter rot disease of grape. Phytopathology 60:1203–1211.

Sutton, B. C., and Gibson, I. A. S. 1977. *Greeneria uvicola*. Descriptions of Pathogenic Fungi and Bacteria, No. 538. Commonwealth Mycological Institute, Kew, Surrey, England.

(Prepared by J. R. McGrew)

White Rot

White rot was first described in Italy in 1878. Its geographical distribution is very similar to that of *V. vinifera*. Because the disease causes the greatest damage in the areas of Europe most prone to hailstorms, it is also known as hail disease. Losses may reach 20–80% of the crop. White rot can also appear in vineyards not hit by hail but subject to summer rain followed by persistent high relative humidity combined with moderate or high temperatures (24–27° C).

Symptoms

Typical symptoms of white rot are found on the clusters. Before *véraison*, a few days after a hailstorm, infected berries assume a yellowish color and later turn pinkish blue. They lose their turgidity and become densely covered with small, brownish violet pustules (immature pycnidia of the causal agent). These structures raise the cuticle from the epidermal layers without rupturing it. Air introduced between the cuticle and the epidermis is responsible for the optical effect that makes the diseased berry appear whitish (Plate 36). Mature pycnidia are grayish white (Plate 37), hence the name white rot.

If conditions are favorable (high temperature and high relative humidity), the disease can spread systemically from an infected berry through the pedicel to the rachis. The result can be the destruction of a major portion of the cluster (Plate 36).

On pedicels disease starts where the pedicels are readily exposed to fungal attack, particularly in loose clusters. The first symptom is a small, pale brown, elongated depression. If temperature and relative humidity remain favorable, the disease progresses over the entire pedicel.

When the fungus spreads systemically into uninjured berries, they gradually turn an increasingly intense pinkish blue color, starting at the pedicel. Under very humid conditions, pycnidia develop on the wrinkled surface of the berry. Under dry conditions, they develop inside the berry on the surface of the seeds. At the end of the season, infected berries fall to the ground, thus providing a source of inoculum for successive years.

If a lesion is located on the main rachis, the portion of the cluster below the lesion dries quickly. Berries on immature infected clusters become pale green and flaccid and later turn brown. Pycnidia do not develop on berries that dry up before the fungus invades them, but they can be found at the site of penetration, provided environmental conditions favor their formation. This aspect of the disease differs from typical white rot and has been mistakenly attributed to physiological drying caused by lack of calcium and magnesium (stem necrosis) or to a water imbalance, for example, rainy periods followed abruptly by drought.

The fungus may also cause cankers on nonlignified shoots. This phase of the disease rarely occurs on *V. vinifera* cultivars, although it is frequently observed in nurseries on American or interspecific hybrid rootstock cultivars. Infections are most prevalent around the nodes of green shoots and appear first as long, depressed, brownish, necrotic areas. The necrotic bark splits, revealing the underlying tissue, which becomes callused. The callus tissue raises the residual fibrous material, which gives the cankers their characteristic frayed look. Cankers on shoots are particularly common in nurseries where vines are allowed to develop freely on the soil surface. Damaged shoots result in a significant loss of wood intended for use as rootstocks and reduce the percentage of successful grafts. Severe damage has been reported on Kober 5BB, 420A, and 3309 C.

The fungus rarely infects leaves.

Causal Organism

Coniella diplodiella (Speg.) Petrak & Sydow (syns. *Coniothyrium diplodiella* (Speg.) Sacc., *Phoma diplodiella* Speg.) is the causal organism of white rot. An ascosporic form has also been described and named *Charrinia diplodiella* Viala & Ravaz, although there is no definite proof that it is part of the life cycle of *Coniella diplodiella.*

The hyphae (12–16 μm wide) have frequent septa and abundant dichotomous branching. They produce appressoria and haustoria. Anastomosis is frequent, and chlamydospores are formed. Pycnidia (Fig. 21) arise from stromata situated under the cuticle. At maturity they are globose and ostiolate,

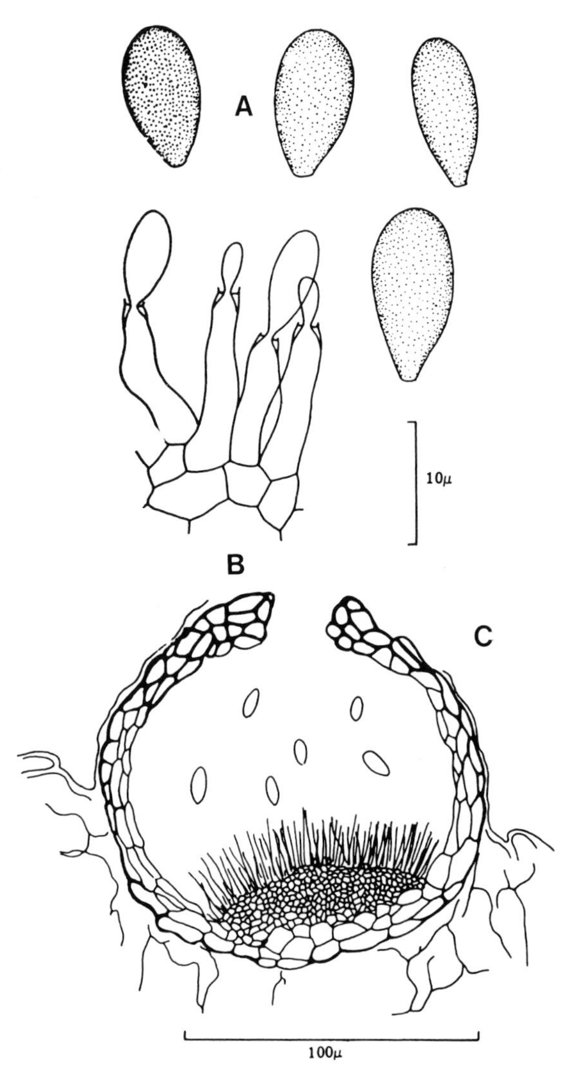

Fig. 21. Conidia (**A**), conidiophores (**B**), and pycnidium (**C**) of *Coniella diplodiella*. (Reprinted, by permission, from Sutton and Waterston, 1966)

measuring 100–150 μm in diameter.

Conidia (Fig. 21) are single-celled and subhyaline to light brown. They measure 8–16 × 5–7 μm and are typically boat-shaped but in many cases are elliptical or ovate, with an obtuse apex and a truncate base. Conidia are immersed in a mucilaginous material and are extruded through the ostiole of the pycnidium.

Disease Cycle and Epidemiology

The life cycle of *C. diplodiella* has two distinct phases: a short parasitic phase and a longer dormant period in vineyard soil.

Clusters become contaminated when particles of soil containing conidia are splashed onto them by rain and hailstones or when farm machinery, such as rotor cultivators, throws soil on the vines. The pathogen is unable to penetrate uninjured berries, and infection takes place only through wounds. Injury caused by hail most frequently favors infection, although clods of soil or stones may also cause injury. Injury caused by other diseases, such as powdery mildew, or insects is less important. The rachis and pedicel, unlike berries, can be infected by direct penetration. Therefore hail, although the principal predisposing factor, is not essential for the occurrence of the disease.

Conidia germinate within a few hours in the juice of injured berries or in raindrops enriched by berry exudates. Conidia germinate and initiate infection rapidly at 24–27° C but slowly at temperatures below 15° C. Disease development is slight above 34° C. If the temperature falls below 15° C for 24–48 hr following a hailstorm, infection is negligible, whereas it may be severe if the temperature remains near 22° C or rises to 24–27° C. The incubation period varies from three to eight days, depending on the tissue infected (longer in shoot penetration), means of penetration, temperature, and relative humidity.

At the end of the season, infected branches, rachises, and desiccated berries fall to the ground, initiating a long dormant period. The pycnidia release thousands of conidia, which remain viable for two to three years. Dried pycnidia can release viable conidia after more than 15 years. White rot-affected vineyards may contain 300–2,000 conidia per gram of soil. Viticultural regions subject to regular hailstorms have the most potential inoculum in the soil, so the frequency and intensity of epidemics in previous years must be considered when forecasting the disease.

Control

Control of white rot is favored by prevention of wounds such as those caused by powdery mildew and insects. Where possible, modify the training system to keep the grape clusters as high above the ground as possible.

Numerous fungicides, such as captan, dichlofluanid, and chlorothalonil, are effective against white rot. Folpet or captan applied within 12–18 hr after a hailstorm provides about 75% control. However, efficacy falls to 50% if treatment is delayed for 21 hr, and treatments after 24 hr are generally ineffective, especially if the temperature is above 20° C. The dicarboximides, if used at temperatures below 20° C and within 12–18 hr after a hailstorm, also provide good control. Soil treatment with high concentrations of captan and thiram, though efficacious against the dormant form of the pathogen, is not economically or ecologically sound.

Selected References

Bisiach, M., and Battino-Viterbo, A. 1973a. Attivitá "in vitro" di alcuni composti chimici contro *Coniothyrium diplodiella*. Not. Mal. Piante 88–89(III S., N. 14–15):73–79.

Bisiach, M., and Battino-Viterbo, A. 1973b. Further researches on grapevine cluster drying-off caused by *Coniothyrium diplodiella*. Meded. Fac. Landbouwwet. Rijksuniv. Gent 38:1561–1571.

Bolay, A. 1963. Le coitre de la vigne. Agric. Romande Ser. B 6:60–62.

Bolay, A. 1977. Le point actuel sur le traitement des vignes par les fongicides après la grele. Prog. Agric. Vitic. 94:233–234.

Locci, R., and Quaroni, S. 1972. Studies on *Coniothyrium diplodiella*. I. Isolation, cultivation and identification of the fungus. Riv. Patol. Veg. 8:59–82.

Sutton, B. C. 1969. Type studies of *Coniella, Anthasthoopa,* and *Cyclodomella*. Can. J. Bot. 47:603–608.

Sutton, B. C., and Waterston, J. M. 1966. *Coniella diplodiella*. Descriptions of Pathogenic Fungi and Bacteria, No. 82. Commonwealth Mycological Institute, Kew, Surrey, England.

(Prepared by M. Bisiach)

Ripe Rot

Ripe rot occurs on grapes as they mature and ripen. This disease was first reported in the United States in 1891 and has since been found in most areas where bunch (*V. labrusca* and *V. vinifera*) and muscadine (*V. rotundifolia*) grapes are grown. Losses vary greatly from region to region and season to season and with susceptibility of the cultivar. Ripe rot has become a serious problem of muscadine grapes in recent years, particularly in the warm, humid areas of the southeastern United States. The fungus also causes a rot of apples, blueberries, and a number of other fruits and vegetables.

Symptoms

The primary symptom of this disease is rotting of the ripe fruit in the vineyard at harvest (Plate 38). Affected berries initially develop circular, reddish brown spots of decay on their skins. The spots subsequently enlarge to include the entire berry. Rotting fruit are characteristically covered with salmon-colored masses of conidia. They may remain attached to the vine or drop as the rot is complete. The berries shrivel as they decay.

Symptoms have not been observed on vegetative portions of grapevines in the United States. However, the causal fungus is reported to cause leaf spots and stem cankers on grapes in The Philippines.

Causal Organism

Colletotrichum gloeosporioides (Penz.) Penz. & Sacc. (teleomorph *Glomerella cingulata* (Stonem.) Spauld. & Schrenk), the cause of ripe rot, produces acervuli subepidermally. The acervuli are arranged in circles. A mass of sticky, salmon-colored conidia is discharged from the acervuli. Conidia are hyaline and guttulate and vary in size (12–21 × 3.5–6 μm) and shape; they are rounded at the ends and are often slightly curved. Perithecia are subspherical and more or less grouped. Asci are subclavate and measure 42–60 × 10–12 μm; ascospores measure 12–24 × 4–6 μm.

Disease Cycle and Epidemiology

C. gloeosporioides survives from one season to another on the vine as dormant mycelium in mummified fruit and infected pedicels. On pedicels, the fungus usually sporulates at the point of berry attachment, often in a ring over the open ends of vascular bundles. Conidia, produced in abundance from these structures during rainy periods in the spring, serve as primary inoculum. Conidia are spread to other parts of the vine by splashing, blowing rain throughout the growing season. Production of conidia from mummies and pedicels is greatest in early spring and decreases during the summer months; few or no spores are produced in August. The primary source of inoculum is also reduced as the mummified fruit detach from the vine and decompose.

Fruit are susceptible to infection by *C. gloeosporioides* at all stages of development, from small green berries to ripe fruit, but they do not show symptoms until ripening. Disease development is favored by warm (25–30° C), wet weather. Conidia germinate, produce appressoria, and penetrate the cuticle of green or ripening fruit within one week under favorable conditions. The fungus then ceases to grow until the fruit mature. Establishment of latent infections is also characteristic of other fruit diseases caused by *C. gloeosporioides*.

Sporulation on ripe fruit near harvest provides a source of secondary inoculum. Frequent rains during this period can result in severe crop loss.

Control

Losses from ripe rot can be reduced by spraying the green berries with protective, broad-spectrum fungicides (e.g., maneb) during the fruit-ripening period. Cultural practices, such as the removal of overwintered mummies from the vine before new growth appears in the spring, also reduce disease development.

Cultivars of *V. rotundifolia* differ greatly in their susceptibility to ripe rot. Resistance of the berry to *C. gloeosporioides* is related to the ability of the vine to prevent fungal development after infection rather than to its ability to prevent initial infection. Dark-skinned grapes (e.g., Nobel, Pride) are resistant, whereas most bronze cultivars (e.g., Carlos, Scuppernong) are susceptible.

Selected References

Clayton, C. N. 1975. Diseases of muscadine and bunch grapes in North Carolina and their control. N.C. Agric. Exp. Stn. Bull. 451. 37 pp.

Daykin, M. E., and Milholland, R. D. 1984a. Ripe rot of muscadine grape caused by *Colletotrichum gloeosporioides* and its control. Phytopathology 74:710–714.

Daykin, M. E., and Milholland, R. D. 1984b. Histopathology of ripe rot caused by *Colletotrichum gloeosporioides* on muscadine grape. Phytopathology 74:1339–1341.

Quimio, T. H., and Quimio, A. J. 1975. Notes of Philippine grape and guava anthracnose. Plant Dis. Rep. 59:221–224.

(Prepared by R. D. Milholland)

Macrophoma Rot

Macrophoma rot affects both bunch grapes (*V. vinifera* and *V. labrusca*) and muscadine grapes (*V. rotundifolia*); however, the disease is more prevalent and destructive on muscadine grapes grown in the southeastern United States. In susceptible muscadine cultivars such as Higgins and Fry, 20–30% of the ripening berries can be lost.

In addition to causing a fruit rot of grape, apple, avocado, and citrus, the causal fungus also has been reported to cause stem blighting and stem cankers on numerous hosts.

Symptoms

One or more circular, flat or slightly sunken lesions, 1–4 mm in diameter, develop on infected berries as they reach maturity (Plate 39). At first the lesions are black, with small tan or buff-colored centers in which pycnidia are embedded. A brown soft rot may develop from the primary lesions and spread over the entire berry in a susceptible cultivar. Affected berries drop from the vines, shrivel, and are finally reduced to dry, hollow shells with pycnidia scattered over the entire surface.

Causal Organism

Botryosphaeria dothidea (Moug. ex Fr.) Ces. & de Not. (syn. *B. ribis* Grossenbacher & Duggar, anamorph *Macrophoma* sp.), the cause of Macrophoma rot, occurs on blighted stems, but only the pycnidial state has been found on diseased berries. Pycnidia are spherical and range from 153 to 197 μm in diameter. Conidia (14–25 × 5–8 μm) are hyaline, unicellular, narrowly elliptical to ovoid, and rounded at either end. Mature ascocarps (172–315 μm in diameter) are ostiolate and spherical. Asci (102–156 × 17–24 μm) are cylindrical and eight-spored and have a thick, two-layered wall. Ascospores (19–31 × 8–11 μm) are hyaline, unicellular, and ovoid to elliptical.

Disease Cycle and Epidemiology

Very little information is available concerning the disease cycle and epidemiology of Macrophoma rot of grape. The optimal temperature for growth and sporulation is about 28° C. The fungus overwinters in the form of pycnidia on infected berries and stems. Pycnidia, spermagonia, and ascocarp initials are produced in botryose stromata on stems during late fall and winter. Conidia are released during wet weather and are disseminated by wind and splashing rain throughout the growing season.

Control

Macrophoma rot can be controlled by applying protective fungicides (e.g., maneb) beginning after bloom and continuing throughout the fruit-ripening period. Cultivars with resistance to this fungus are available. The muscadine cultivars Hunt and Scuppernong are resistant, whereas Chowan, Higgins, and Fry are susceptible. Carlos and Magnolia are intermediate in susceptibility.

Selected References

Clayton, C. N. 1975. Diseases of muscadine and bunch grapes in North Carolina and their control. N.C. Agric. Exp. Stn. Bull. 451. 37 pp.

Jenkins, W. A. 1941. Diseases of muscadine grapes. Pages 19–29 in: Further Studies with Muscadine Grapes. L. Ascham, T. L. Bissell, W. L. Brown, T. A. Pickett, E. F. Savage, M. M. Murphy, Jr., and J. G. Woodroof, eds. Ga. Agric. Exp. Stn. Bull. 217.

Luttrell, E. S. 1948. *Botryosphaeria ribis*, perfect stage of the *Macrophoma* causing ripe rot of muscadine grapes. Phytopathology 38:261–263.

(Prepared by R. D. Milholland)

Angular Leaf Spot

Angular leaf spot is an important disease of muscadine grapes (*V. rotundifolia*) in most vineyards throughout the southeastern United States. Only the leaves of muscadine grape are affected. The primary damage to susceptible cultivars is premature defoliation, which results in reduced plant vigor and yield.

Symptoms

Lesions first appear on the adaxial leaf surface as small chlorotic flecks. As lesions develop, a small, dark brown to black area appears in the center and is surrounded by a distinct halo (Plate 40). Halos are indistinct or absent on the abaxial leaf surface. Lesions become angular to irregular in shape and vary from one to several centimeters in diameter two months after infection. The necrotic areas appear papillate under magnification because of numerous tufts of conidiophores.

Causal Organism

Mycosphaerella angulata Jenkins (anamorph *Cercospora brachypus* Ell. & Ev.) is the causal organism of angular leaf spot. Conidia (16.8–112 × 2.2–3.5 μm) are hyaline, slender, curved, and more or less acute at each end, with one to five septations. The number of septations is influenced by the relative humidity. Spermatia (2–4 × 0.5–0.7 μm) are uninucleate and rod-shaped at maturity. Perithecia (40–90 × 40–60 μm) are ostiolate, black, ovate, amphigenous, and partly embedded in overwintered host leaves. Ascospores (14–19.6 × 2.8–5.6 μm) are bicellular, straight to slightly curved, hyaline, and guttulate.

Disease Cycle and Epidemiology

Numerous light olive-colored spores are produced on both leaf surfaces under humid conditions but are more abundant on

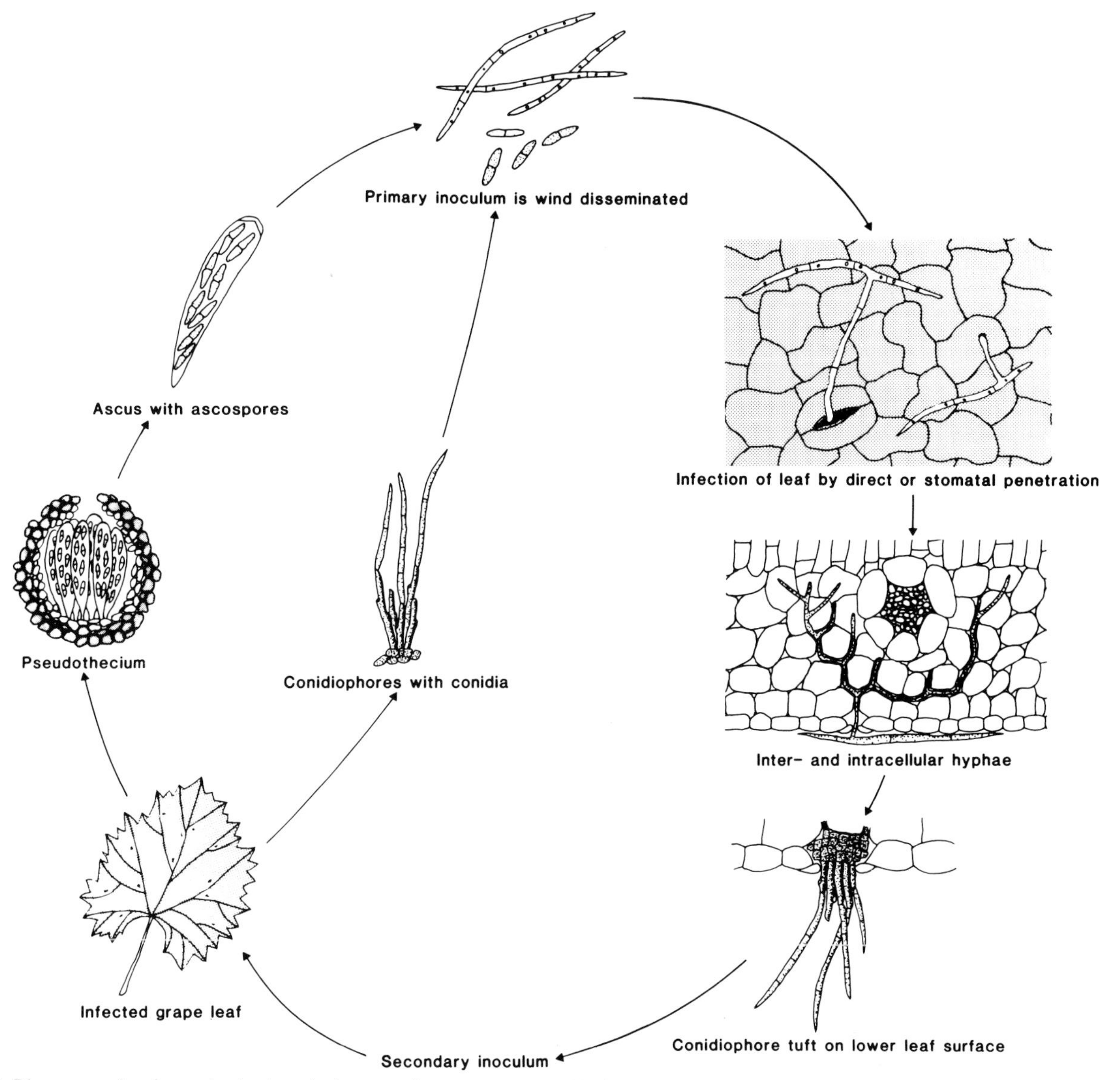

Fig. 22. Disease cycle of angular leaf spot of muscadine grape, caused by *Mycosphaerella angulata.* (Drawing by M. E. Daykin)

the abaxial surface. Infection by conidia or ascospores occurs through either leaf surface by direct penetration or indirectly through stomata (Fig. 22). Hypertrophy of spongy mesophyll cells is associated with intercellular growth of the fungus in early stages of infection. Intracellular colonization occurs in advanced infections, leading to necrosis and collapse of cells.

Tufts of conidiophores develop from subepidermal hyphae on the abaxial leaf surface and produce conidia shortly after they emerge. Spermagonia originate from either leaf surface and mature in October and early November. Various developmental stages may be found in early spring on the overwintered leaves. Perithecial development is initiated concurrently with the spermagonia. Mature perithecia develop from overwintered leaves in the spring, and after a wetting and drying period, ascospores are discharged. Both ascospores and conidia serve as primary inoculum and are disseminated by wind.

Control

Angular leaf spot can be controlled with applications of protective fungicides starting after bloom and repeated at 14-day intervals until late August. Muscadine cultivars vary in susceptibility to angular leaf spot; Magnolia and Scuppernong are more susceptible than Chowan or Carlos.

Selected References

Clayton, C. N. 1975. Diseases of muscadine and bunch grapes in North Carolina and their control. N.C. Agric. Exp. Stn. Bull. 451. 37 pp.

Ellis, J. B., and Everhart, B. M. 1902. New Alabama fungi. J. Mycol. 8:62–73.

Jenkins, W. A. 1942. Angular leaf spot of muscadines, caused by *Mycosphaerella angulata* n. sp. Phytopathology 32:71–80.

(Prepared by R. D. Milholland)

Diplodia Cane Dieback and Bunch Rot

Diplodia cane dieback kills canes, spurs, arms, and trunk wood. The causal fungus may also initiate a fruit rot in some vineyards. The disease has been observed in *V. vinifera* cultivars Black Monukka, Cardinal, Delight, Emperor, Italia, Perlette, Ribier, and Thompson Seedless. The disease has been reported in Egypt, Israel, and India and in Arizona and California in the United States.

Symptoms

Infected canes typically die back from the tip toward the base (Plate 41), starting in summer. Dead parts are brown to gray and are speckled with black dots (pycnidia). Cankers that form on canes at the base of a diseased cluster stem and also those in girdle wounds in canes and trunks may advance in any direction.

In fall, winter, and spring, black pycnidia show in the bark of diseased canes and spurs and under bark of cankers on arms and trunks. Spurs or portions of spurs and arms may die. In summer, pycnidia erupt through new bark, in cracks, and under old bark scales of diseased parts.

Berries affected by Diplodia bunch rot at first appear slightly water-soaked and in cultivars with white berries become a light pink. As the rot progresses, the skin cracks and the berries drip juice and become covered with a cottony white mass of mycelium. In the absence of other organisms, infected berries dry out and turn to leathery mummies, with black pycnidia emerging in dark ridges. However, this form of Diplodia rot is seldom observed in the vineyard because the juice from infected berries attracts vinegar flies and other insects, which introduce spores of fungi and yeasts. Soon, these organisms turn the inside of the cluster into a teeming, rotten, juicy, vinegary smelling mass, a condition known as summer bunch rot.

Causal Organism

The causal organism is *Diplodia natalensis* Pole Evans (syns. *Botryodiplodia theobromae* Pat., *D. viticola* Desm.). *Botryosphaeria rhodina* von Arx (syn. *Physalospora rhodina* (Berk. & Curt.) Cooke) is reported to be the ascogenous stage but is seldom observed and does not seem to be involved in the common disease.

The mycelium is light gray to black, with or without gray to black stromata. Pycnidia are black, globose, beaked, subcutaneous to erumpent, singular, and scattered or in clumps (sometimes fused); they occur either in or on a stroma and are hirsute or glabrous, with paraphyses present when young and absent when mature. Pycnidia vary from 93 to 625 μm in diameter.

Conidia (24 × 14 μm) are extruded through the beaks of pycnidia in a wet, gelatinous matrix or in dry, black cirri. Young conidia are hyaline, granular, and one-celled. Mature conidia are dark and have rounded ends, conspicuous longitudinal striations, and one central septum.

The temperature range for growth of *D. natalensis* is 9–39° C; the growth rate is maximum between 27 and 33° C.

Disease Cycle and Epidemiology

The fungus overwinters in diseased parts of canes, spurs, arms, and trunks of vines and on vine parts in or on the soil. In spring, as air and soil temperatures increase and vine growth becomes vigorous, *D. natalensis* resumes activity.

The fungus becomes soilborne from diseased prunings either on or in the soil and has been recovered from vineyard soil samples taken from depths of 1 m. Large numbers of conidia are exuded in cirri from pycnidia on diseased vine parts in moist to wet soil. Conidia are commonly windblown in vineyard dust and may be waterborne in splashed droplets from rain or sprinkler irrigation.

Tips of shoots may become infected when they touch wet soil or when covered by soil during cultivation. Infections in the green tips move rapidly up the shoot, eventually causing the cane dieback disease. Some of the shoot infections may grow into the spur and arm, killing tissues as they advance.

Berries become infected during or shortly after bloom from conidia that are windblown in vineyard dust. A germ tube penetrates the stigma, forms a bit of mycelium, and becomes latent in the tissue at the base of the stigma. Many if not most of the infections abort or are walled off in the stigma scar. However, as some infected berries mature to 12–15% sugar, the fungus resumes growth and rots the berry. The fungus may move from berry to berry in a cluster, resulting in bunch rot. The pathogen may also invade the cluster stem and grow into the cane, causing a canker.

Diplodia cane dieback and bunch rot is a disease of warm to hot weather. It has been most damaging to vineyards in hot temperate to semitropical and desertlike regions. The disease is favored by high relative humidity, occasional summer rains, furrow or sprinkler irrigation, high inoculum levels, and cultural conditions conducive to rapid vine growth. *D. natalensis* is endemic in many vineyards in warm to hot regions of California; it is also present in some cool areas but goes unnoticed at a low level, affecting only a few spurs and arms.

Control

A low level of disease can be tolerated by keeping the inoculum level in the vineyard as low as possible. To minimize inoculum, prune all diseased wood during pruning time, prune and train to prevent cluster clumping, remove prunings from the vineyard and destroy them, and do not cultivate or create dust in the vineyard during the time of bloom to shatter or when girdle wounds are open. Mowing off tips of shoots about 15 cm above the ground before cultivation to prevent shoot contact with soil is also recommended.

Selected References

El-Goorani, M. A., and Maleigi, M. A. 1972. Dieback of grapevine by *Botryodiplodia theobromae* Pat. in Egypt. Phytopathol. Mediterr. 11:210–211.

Hewitt, W. B. 1974. Rots and bunch rots of grapes. Calif. Agric. Exp. Stn. Bull. 868. 52 pp.

Leavitt, G. M., and Munnecke, D. E. 1987. The occurrence, distribution, and control of *Botryodiplodia theobromae* on grapes (*Vitis vinifera*) in California. (Abstr.) Phytopathology 77:1690.

Patil, L. K., and Moniz, L. A. 1969. A new anthracnose disease of grapevine from India. J. Univ. Poona Sci. Technol. 36:107–110.

Punithalingam, E. 1976. *Botryodiplodia theobromae.* Descriptions of Pathogenic Fungi and Bacteria, No. 519. Commonwealth Mycological Institute, Kew, Surrey, England.

Strobel, G. A., and Hewitt, W. B. 1964. Time of infection and latency of *Diplodia viticola* in *Vitis vinifera* var. Thompson Seedless. Phytopathology 54:636–639.

Webster, R. K., Hewitt, W. B., and Polach, F. J. 1969. Studies on *Diplodia* and *Diplodia*-like fungi. III. Variation in *Diplodia natalensis* from grape in California. Hilgardia 39:655–671.

Webster, R. K., Hewitt, W. B., and Bolstad, J. 1974. Studies on *Diplodia* and *Diplodia*-like fungi. VII. Criteria for classification. Hilgardia 42:451–463.

(Prepared by W. B. Hewitt)

Berry Rots and Raisin Molds

Miscellaneous Berry Rots and Bunch Rots

Berry rots and bunch rots are common viticultural problems the world over. Many rot organisms infect individual berries and seldom spread to others in the cluster, whereas other rot organisms typically spread to many or most of the berries in the cluster, causing bunch rots. Some organisms cause storage and transit rots that, although mostly initiated in the vineyard, develop after harvest.

Annual losses to berry rots are typically 3–5% but in unfavorable seasons may range from 15 to 80%. With few exceptions, rotting berries are not suitable for making wine or other products such as vinegar because they impart odd and distasteful flavors.

Various species of fungi, including yeast, and some bacteria contribute to the rotting of berries. Some 70 species of fungi in 30 genera and also a few bacteria, including acetic acid species, have been found on berries, and many of these organisms are directly or indirectly involved in rot. Some infect green berries, but infections do not necessarily advance into rot. Others produce unsightly spots or scars on green berries, and still others become latent and later advance into rot as berries mature. However, many of these 70 species seem to be transient, occurring in abundance at specific times but not necessarily building up during the rot season.

In general, the fruit-rotting organisms may be divided into two major groups: those that infect berries directly, or "primary invaders," and those that enter berries through wounds or follow a primary invader, known as "wound and secondary invaders." Wounds commonly found on berries are cracks in the skin caused by internal pressure associated with rain or due to diseases such as powdery mildew and black measles. Hail, bird pecks, and insect feeding also cause wounds that permit entry of secondary invaders.

Evidence of rot problems in vineyards includes the odor of vinegar, juice dripping over berries of a cluster, the presence of flies and larvae of the vinegar fly (*Drosophila melanogaster* Meigen), soft spots in the skin of berries ("slip-skin"), and fungus fruiting structures on the surface of berries.

Some specific rot diseases are described in the following subsections under the genus name of the causal organism, which is commonly used to name a specific rot.

Primary Invaders. Alternaria rot caused by *Alternaria alternata* (Fr.) Keissl. (syn. *A. tenuis* Nees) often occurs near the cap stem. The rot area is tan at first and becomes brown with age. Under humid conditions, fluffy gray tufts of fungus and conidiophores bearing conidia develop through cracks in the brown skin over rot lesions. Infections may occur through the skin in a small drop of water or at 98–100% relative humidity. Rot at the cap stem may also develop from an infected pedicel. Infections of pedicels occur during most seasons. The fungus has been cultured from pedicels during the time of blossom shatter to berry maturity.

Cladosporium rot caused by *Cladosporium herbarum* (Pers.:Fr.) Link is a well-defined, black, soft, circular area ranging from 5–7 mm in diameter up to as much as two-thirds of a berry. At room temperature, in a humid atmosphere, the surface of the rot area becomes a velvet olivaceous color because of the presence of conidiophores and conidia. Cladosporium rot is typically a storage disease that develops on fruit harvested late in the season after a rain. Infection takes place through the skin either in the vineyard or in storage and may occur at temperatures from 4 to 30° C (optimum 20–24° C).

Other primary invaders include *Botrytis cinerea* Pers., *Diplodia natalensis* Pole Evans, *Elsinoë ampelina* (de Bary) Shear, *Glomerella cingulata* (Stonem.) Spauld. & Schrenk, *Guignardia bidwellii* (Ellis) Viala & Ravaz, *Greeneria uvicola* (Berk. & Curt.) Punithalingam, and *Phomopsis viticola* (Sacc.) Sacc. They are covered elsewhere in this compendium.

Wound and Secondary Invaders. Aspergillus rot caused by *Aspergillus niger* van Tiegh. is common in warm to hot climates. This rot is usually associated with a wound and is at first tan to brown, but the area is soon covered with a dusty mass of brown or black spores. These rot areas are initially soft but become firm and leathery. Under warm conditions (20–32° C) in a drop of water, the fungus may infect mature fruit directly through the skin.

Blue mold or Penicillium rot caused by *Penicillium* spp. is common in wounded berries (Plate 42). The fungus produces large, dusty masses of colored spores.

Rhizopus rot is a wet, juicy rot caused by *Rhizopus arrhizus* Tesher or *R. stolonifer* (Ehrenb.:Fr.) Lind in warm, humid climates. Rotted areas on berries are soft and brown, drip juice, and under humid conditions may be covered with a cobwebby mycelium. Numerous sporangiophores with small, spherical, dark sporangia emerge through cracks in the skin of diseased berries or through the borders of a wound. In humid weather the fungus may spread to other berries in a cluster, causing a bunch rot (Plate 43). The disease is often abundant in vineyards near plum and peach trees with decaying fruit on the ground and also near fields of sugar beets.

Fruit rots caused by *Ascochyta* sp., *Fusarium moniliforme* Sheldon, *Hormiscium* sp., *Stemphylium botryosum* Walker, and *Torula* sp. are indiscrete rots or include a mix of organisms. Rot caused by *Helminthosporium* sp. alone appears much like that caused by *Cladosporium herbarum.*

Alternaria geophila Deszew., *Aspergillus niger, R. arrhizus,* and *R. stolonifer* become primary invaders when provided with a drops of water or grape juice in the temperature range of 18–30° C. A drop of water on the skin takes up sugars and amino acids from the berry and provides a good food base for the pathogens.

Sour bunch rots (Plate 44) in general are caused by a mix of various fungi, yeasts, acetic acid bacteria, larvae of the fruit fly, and other organisms. Affected berries drip juice and smell like vinegar. Sour bunch rots may have different causes, but many are initiated by the rot of one or two injured berries in a cluster. Juice from the rotting berries that drips onto other berries may induce cracks in the skin of mature fruit or serve as a medium for the growth of other rot fungi.

Other wound and secondary invaders include fungi such as *Aspergillus aculeatus* Iizuka, *A. flavus* Link, *A. ochraceus* Wilhelm, *A. wentii* Wehmer, *Botryosphaeria dothidea* (G. & P.) Arx & Müller, *Candida* sp., *Chaetomium elatum* Kunze, *Cladosporium cladosporioides* (Fres.) de Vries, *C. oxysporium* Berk. & Curt., *Monilia* sp., *Penicillium brevicompactum* Diercks., *P. cyclopium* West., *P. frequentaus* West., *P. stoloniferum* Thom, *Saccharomyces cerevisiae* Kreger-van Rij, and *Sclerotinia sclerotiorum* (Lib.) de Bary.

Disease Cycle and Epidemiology. Many fruit-rotting organisms are fungi that produce large numbers of spores that are windblown, carried in dust, spread in rain, or transported on feet and mouthparts of insects. These fungi occur in vineyards on mummified berries, dead vine parts, bark of canes and spurs, vineyard debris, and other decaying vegetation.

Species of *Alternaria, Cladosporium,* and *Stemphylium* grow in bark of one-year-old canes and sporulate there in the spring. These three fungi, in addition to *Botrytis* and *Helminthosporium,* grow under humid conditions and sporulate on cast-off flower parts that remain in the cluster.

Berry rots and bunch rots are wet-weather disease problems. They are favored by high relative humidity, rain, and sprinkler irrigation during the time berries are maturing and approaching harvest. The longer the wet period, the greater the amount of rot. Rots become increasingly abundant each year in a sequence of favorable seasons but may stop abruptly in one dry season. Some cultivars are more prone to rot than are others.

Control. Several viticultural practices may aid in reducing rot. Prune to adjust the crop for early maturity and prevent clumping of clusters. Thin if necessary to loosen clusters. Clean the vineyard and surrounding area of debris, prunings, and unnecessary vegetation. Adjust time and duration of sprinkler irrigation so that berries are not kept wet more than 18 hr, and do not sprinkle after fruit reaches 15% sugar. Keep the vine open so that fruit is well aerated. Control agents that injure fruit, such as the powdery mildew fungus, birds, and insects.

Raisin Molds and Rots

Molds and rots of raisins occur periodically in association with wet weather. The amount of mold and rot varies with the vineyard, the time of harvest, the district, and the season. Losses range from 0 to 2% in a normal season and may be as high as 70% in a rainy season. These diseases reduce production, increase the cost of making and processing raisins, and reduce quality. The amount of preharvest berry rot in a vineyard grown for raisins is an indication of the potential for mold and rot of raisins while drying.

Symptoms. Molds are mostly contaminants. Mold fungi grow on the surface of raisins, where they produce conidial spore masses. They can be cleaned (brushed or washed) from the surface. Rots are caused by fungi that colonize the inside of the berry before harvest or during drying. Fungi growing inside the raisin form mats of mycelium and sporulation tufts on the surface. The internal mycelial mats can be observed after infected raisins are soaked in a relatively large volume of warm water or are boiled in water until clear. Raisins with rot are a total loss because the rot cannot be removed by washing.

Causal Organisms. The same fungi found normally in vineyards are also found on berries when harvested and laid to dry on trays in the vineyard or on racks or drying pads. Species of *Alternaria, Aspergillus, Botrytis, Chaetomium, Cladosporium, Helminthosporium, Hormiscium, Hormodendrum, Penicillium, Rhizopus,* and *Stemphylium* are common raisin mold and rot fungi.

Disease Cycle and Epidemiology. At harvest, berries have on their skin a representative sample of spores of fungi that occur naturally in the vineyard. When berries are moistened by dew or rain, sugars and amino acids of the berry move into surface water. Conidia germinate and grow in this medium on the surface of berries. Within a few hours, these colonies sporulate, and a new crop of conidia may be blown by wind or splashed by raindrops to spread the organism.

Berries that have rot at the time of harvest and are laid on trays to dry also support fungal sporulation. Conidia on these berries are also spread by wind or rain to other berries. Furthermore, when water remains on the surface of berries or on drying trays in contact with berries for 24–36 hr or more, some fungi, such as species of *Alternaria, Aspergillus, Cladosporium,* and *Rhizopus,* penetrate the skin and colonize the interior of the berry or raisin and cause rot. These fungi may also grow on the paper tray, causing the raisins to stick to the paper. Rains or showers that keep the fruit wet for periods of 24–30 hr or longer at temperatures of 15–23° C produce ideal conditions for the development of molds and rots.

Control. Two approaches may be taken to control raisin molds and rots: prevention and chemical treatments. Although molds and rots on raisins generally occur after harvest while the fruit is on drying trays in the vineyard, preventive control begins early in the season and continues through drying and curing. Chemical control is a late choice and often a futile effort.

To prevent preharvest rot in the vineyard, follow suggested controls for berry and bunch rots. Adjust the crop to obtain high soluble solids early in the harvest and drying season to maximize quality and permit flexibility in choice of harvesting time. Choose harvesttime in accordance with the best weather forecast for satisfactory drying. Cultivate between vine rows to reduce debris. Prepare a good tray base by forming soil to a firm slope, faced to receive maximum sun exposure and good drainage in case of rain. Harvest with care to minimize damage to fruit; that is, *cut* clusters from the vine and *place* them in only one layer on the drying tray. Turn the fruit over when it is partially dry to shorten the drying time.

The effectiveness of a fungicide to prevent mold and rot of raisins depends on the amount of inoculum, the amount and degree of damage to berries at harvest, the fungicide used, the timing and thoroughness of fungicide application, and the duration of the wetting period. Fungicides have given reasonable protection when applied to the fruit before harvest or to fruit on drying trays soon after harvest but before rain.

Selected References

Barbe, G. D., and Hewitt, W. B. 1965. The principal fungus in the summer bunch rot of grapes. Phytopathology 55:815–816.

Bisiach, M., Minervini, G., and Salomone, M. C. 1982. Recherches expérimentales sur la pourriture acide de la grappe et sur ses rapports avec la pourriture gris. Bull. OEPP/EPPO Bull. 12:15–27.

Delp, C. J., Hewitt, W. B., and Nelson, K. E. 1951. Cladosporium rot of grapes in storage. (Abstr.) Phytopathology 41:937–938.

Harvey, J. M., and Pentzer, W. T. 1960. Market diseases of grapes and other small fruits. U.S. Dep. Agric. Agric. Handb. 189. 37 pp.

Hewitt, W. B. 1974. Rots and bunch rots of grapes. Calif. Agric. Exp. Stn. Bull. 868. 52 pp.

Martini, L. P. 1966. The mold complex of Napa Valley grapes. Am. J. Enol. Vitic. 17:87–94.

Nelson, K. E., and Ough, C. S. 1966. Chemical and sensory effects of microorganisms on grape musts and wine. Am. J. Enol. Vitic. 17:38–47.

(Prepared by W. B. Hewitt)

Rust

Grape rust occurs in the tropics, extending into the North Temperate Zone of Asia from Sri Lanka, India, and Java northward to Korea and Japan, and in the Americas from Colombia, Venezuela, and Central America through the West Indies to southern Florida in the United States and occasionally north of the frost line in the southern United States. The disease is common in Asia and Central America and can be very destructive if not controlled.

Symptoms

Small, yellowing pustules of uredia, either scattered or densely distributed, appear on the lower leaf surface (Plate 45) and occasionally on petioles, young shoots, and rachises. On some cultivars, brown necrotic spots develop on the upper leaf surface opposite the uredial pustules. The lesions appear primarily on the mature leaves. Severe infection causes premature defoliation and reduced growth in the current season. In later stages of disease development, telia appear as brown to dark brown protrusions near or within the uredia (Plate 46).

Causal Organism

Physopella ampelopsidis (Diet. & Syd.) Cumm. & Ramachar (syns. *Phakopsora ampelopsidis* Diet. & Syd., *Angiopsora ampelopsidis* (Diet. & Syd.) Thirum. & Kern, *Uredo vitis* Thüm., *U. vialae* Lagerh., *Physopella vitis* Arth.), the cause of grape rust, is a macrocyclic rust that produces pycnia (Plate 47) and aecia (Plate 48) on *Meliosma myriantha,* a deciduous tree in Japan. Pycnia are nearly round, measure 100–130 μm in diameter, are brown to black, and protrude from the upper leaf surface. Aecia protrude from the lower leaf surface and are 150–200 μm in diameter. Peridial cells are compactly arranged, with smooth outer walls 5–7 μm thick and spinelike inner walls 4–13 μm thick. Aeciospores are ovoid to obovate (15–20 × 12–16 μm), have fine spines, and are colorless and single-celled. This rust fungus on *M. myriantha* is known as *Aecidium meliosmae-myrianthae* P. Henn. & Shirai.

The uredial and telial stages are produced on *Vitis* and *Ampelopsis.* Uredia appear as yellowish pustules, 0.1–0.5 mm in diameter, on the lower leaf surface. Uredospores are broadly ellipsoid to obovate (18–29 × 10–18 μm), with walls almost colorless to pale yellow and 1.5 μm thick; they are minutely but closely echinulated, and pores are indistinct. Paraphyses are cylindrical (30–70 × 6–11 μm), numerous, curved and irregular, and yellowish. Telia are hypophyllous, scattered, roundish, 0.1–0.2 mm in diameter, and three to six cells thick. Teliospores form in chains, are ovoid (10–35 × 9–15 μm), have smooth walls, and are nearly colorless.

Other rust fungi recorded on *Vitis* include *Phakopsora cronartiiformis* (Barcl.) Diet., which has cinnamon brown, verruculose uredospores and teliospores in crusts of laterally adherent spores; *Catenulopsora vitis* (Butl.) Mund. & Thirum. (syns. *Chrysomyxa vitis* Butl., *Kuehneola vitis* (Butl.) Syd.), which has powdery telia with teliospores in tenacious chains, with spores joined laterally; and *U. caucensis* Mayor, reported from Colombia, which has slightly larger uredospores (21–30 × 16–22 μm) than *Physopella ampelopsidis.* Two species of *Aecidium, A. vitis* Smith and *A. guttatum* Kunze, have also been recorded on *Vitis.*

Disease Cycle and Epidemiology

Basidiospores from germinated teliospores infect *M. myriantha,* producing pycnia and later aecia on the lower surface of leaves (Fig. 23). The aeciospores infect *Vitis.* Pycnia and aecia have been reported only from Japan.

In most other areas, only uredia and telia are produced (Fig. 23). Uredia can be found in tropical and subtropical areas almost year-round. Telia usually develop when the weather turns cooler, appearing in late autumn in temperate regions, but they can be found as early as July in Taiwan. In tropical and subtropical areas, the fungus overwinters on the vine as uredospores on green tissue.

Grape rust is more severe in tropical and subtropical areas than in temperate regions. Cardinal temperatures for uredospore germination in water are 8, 24, and 32° C (minimum, optimum, and maximum, respectively). High humidity enhances germination, but light is detrimental. Appressoria can be observed 6 hr after inoculation and penetrate through stomatal openings after 12 hr. Five days after inoculation, the colonized area appears approximately 200–300 μm in diameter. Uredial sori with uredospores develop after seven days and are 300–400 μm in diameter. Pustules appear five to six days after inoculation at 16–30° C and after 15–20 days at 12° C. The uredospores do not infect young leaves in which the stomata are not well developed. Teliospores germinate at 10–30° C; temperatures of 15–25° C are optimal. Basidiospore formation is optimal between 15 and 25° C; germination occurs at 5–30° C and is optimal at 20–25° C. High humidity at night is necessary for development of epidemics.

Control

Cultivars derived from *V. labrusca, V. vinifera, V. aestivalis,* and most other temperate species are susceptible to grape rust, whereas those derived from the tropical group, *V. tiliaefolia, V. simpsoni,* and *V. coriacea,* are almost immune.

Fungicides such as Bordeaux mixture, zineb, maneb, ferbam, and captafol are effective against grape rust. Fungicide spraying begins when a light incidence of rust occurs and is repeated at intervals of 10–14 days.

Selected References

Clayton, C. N., and Ridings, W. H. 1970. Grape rust, *Physopella ampelopsidis,* on *Vitis rotundifolia* in North Carolina. Phytopathology 60:1022–1023.

CMI. 1985. *Physopella ampelopsidis* (Dietel & Sydow) Cumm. & Ramachar. 4th ed. Distribution Maps of Plant Diseases, No. 87. Commonwealth Mycological Institute, Kew, Surrey, England.

Fennell, J. L. 1948. Inheritance studies with the tropical grape. J. Hered. 39:54–64.

Kuro, A., and Kaneko, S. 1978. Heteroecious nature of grape rust fungus. (Abstr.) Ann. Phytopathol. Soc. Jpn. 44:375. (In Japanese)

Leu, L. S., and Wu, H. G. 1983. Uredospore germination, infection and colonization of grape rust fungus, *Phakopsora ampelopsidis.* Plant Prot. Bull. (Taiwan) 25:167–175. (In Chinese, English summary)

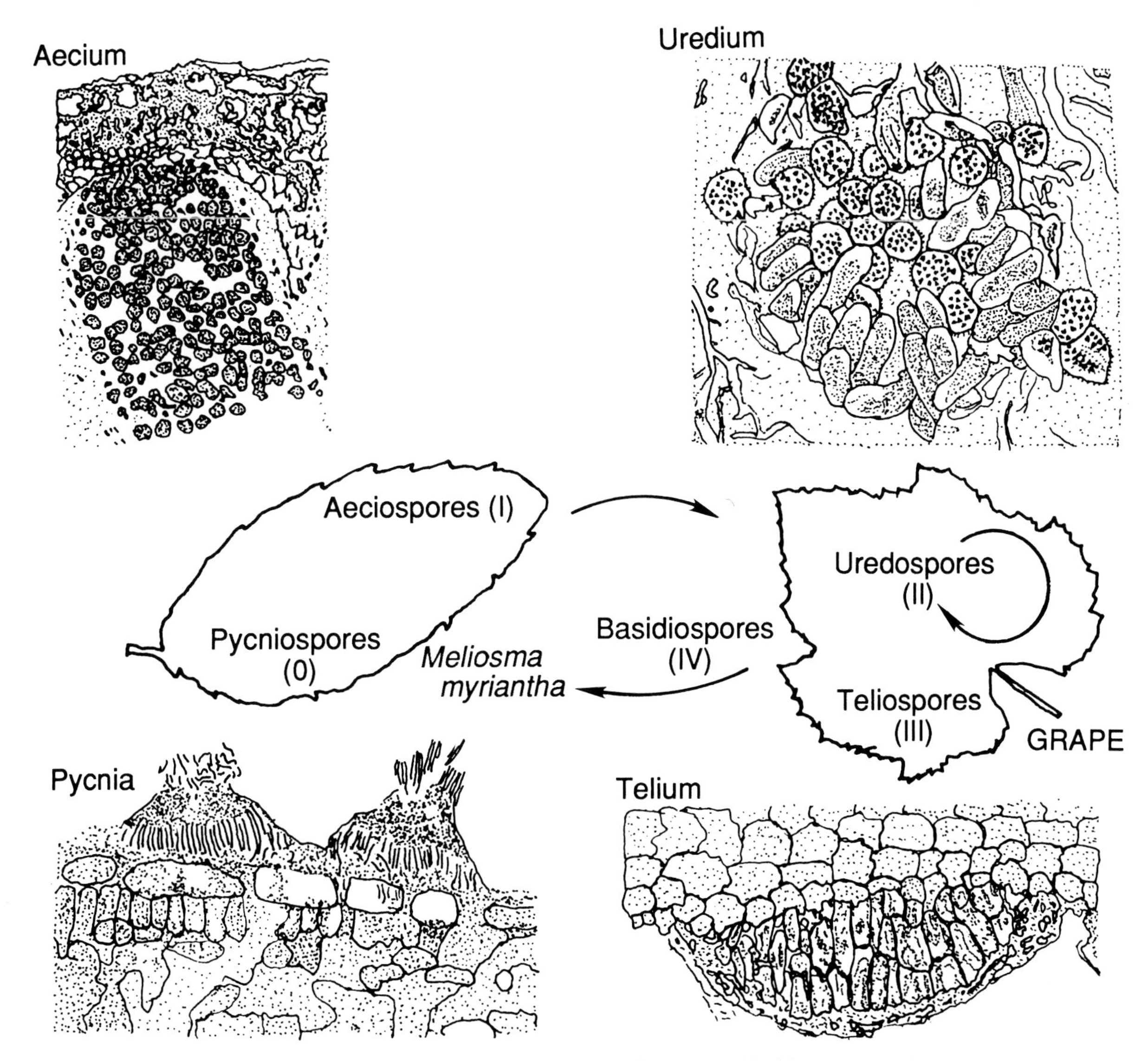

Fig. 23. Disease cycle of grape rust, caused by *Physopella ampelopsidis.* (Courtesy A. Kudo)

Punithalingam, E. 1968. *Physopella ampelopsidis*. Descriptions of Pathogenic Fungi and Bacteria, No. 173. Commonwealth Mycological Institute, Kew, Surrey, England.

(Prepared by L. S. Leu)

Minor Foliage Diseases

The literature contains a formidable number of reports of fungi described on grape foliage. Most of these fungi were collected from vines of the many wild species of *Vitis* in diverse climates. Many are localized or limited to a few host species, some have a wide host range that happens to include *Vitis*, and all appear under a multitude of synonyms.

Minor foliage diseases are generally not a problem in vineyards that receive general disease control practices, but some may occasionally bridge the gap between wild hosts and cultivated vines when conditions favor their growth. They tend to be found in newly set vineyards and are sometimes reported by alarmed novice growers who cannot identify them as one of the usual diseases described in reference manuals.

Leaf Blight

Leaf blight, also called Isariopsis leaf spot, occurs primarily in the southeastern United States, although it has been reported from Massachusetts, Connecticut, Kansas, Illinois, and California on vines of wild species. It has been reported throughout the warmer grape-growing areas of the world under one or another of its several synonyms. It is not reported on muscadine grapes. The disease tends to appear after harvest when spraying is discontinued and may cause considerable defoliation in a wet year if not controlled.

Symptoms

Spots are irregular to angular, sometimes with a serpentine outline. They are brown, measure 2–20 mm in diameter, and coalesce. They have clearly defined borders on the upper leaf surface and diffuse margins on the lower leaf surface. They appear first on the lower, shaded leaves. Lesions soon become black and brittle.

Causal Organism

Pseudocercospora vitis (Lév.) Speg. (syn. *Isariopsis clavispora* (Berk. & Curt.) Sacc.), the imperfect stage of *Mycosphaerella personata* Higgins, is the cause of leaf blight. The fruiting structures are slender, black, bristlelike synnemata (200–500 μm long) bearing olive brown, elongate conidia (25–99 × 4–8 μm) with three to 17 septa.

The perfect stage may develop on dead leaves late in the season. Perithecia are round (60–90 μm in diameter), black, embedded, and warty above. Asci are club-shaped (30–40 × 6–10 μm), and ascospores (10–20 × 2.5–3.6 μm) may be forcibly discharged.

Selected References

Deighton, F. C. 1976. Studies on *Cercospora* and Allied Genera. VI. *Pseudocercospora* Speg., *Pantospora* Cif. and *Cercoseptoria* Petr. Mycological Paper 140. Commonwealth Mycological Institute, Kew, Surrey, England. 168 pp.

Higgins, B. B. 1929. Morphology and Life History of Some Ascomycetes with Special Reference to the Presence and Function of Spermatia. II. Ga. Agric. Exp. Stn. J. Ser. Pap. 28. Pages 287–296.

Rhoads, A. S. 1926. Diseases of grapes in Florida. Fla. Agric. Exp. Stn. Bull. 178:123–125.

(Prepared by J. R. McGrew and F. G. Pollack)

Leaf Blotch

The fungus that causes leaf blotch is widespread throughout the eastern United States (New York to Wisconsin and south to Texas and North Carolina) and has been reported from Italy and northern Portugal. Lesions are most often found on the foliage of rootstock cultivars derived from American *Vitis* spp. but sometimes appear on *V. labrusca, V. vinifera,* and *Vitis* interspecific hybrid cultivars. The fungus has been found fruiting on berries that were left on the vine past normal harvest. Only fruit infection has been reported from Italy.

The leaf blotch fungus has been found in association with the larvae of the grape leaffolder, *Desmia funeralis* (Hübner), fruiting on frass inside the rolled-up portion of the leaf blade and sometimes producing a leaf lesion with synnemata in the immediate area of the infested frass.

Inoculation trials in Portugal indicate that rootstock cultivars are more susceptible than *V. vinifera* and interspecific hybrid fruiting cultivars. Nevertheless, in Maryland, even when rootstocks were grown without fungicidal sprays and leaf blotch was obviously present, the overall reduction in effective leaf area was minor.

Symptoms

Leaf lesions generally appear after midseason and range in size from 5 mm in diameter to as much as one-third of a leaf (Plate 49). Smaller lesions may have distinct, dark margins, while larger ones show distinct, light-colored zonate rings or arcs. Synnemata are produced within three or four days of the appearance of a lesion.

Causal Organism

Briosia ampelophaga Cav., the cause of leaf blotch, forms scattered and conspicuous synnemata, more frequently on the lower but sometimes on both surfaces of leaf lesions. Synnemata have relatively thick white stipes consisting of parallel hyaline hyphae that are massed, anastomosed, and interwoven to form rigid upright structures as long as 1 mm. The stipe is capped with a dark, spherical mass, which may be 1 mm in diameter, consisting of chains of dry, dark brown conidia. Individual conidia are globose and measure 3–5 μm in diameter.

Selected References

Cavara, F. 1888. Intorno al dissecamento dei grappoli della vite. Atti Ist. Bot. Univ. Pavia (Ser. 2) 1:321.

Doutel Serafim, F. J. 1955. *Coremium luteolum* S. Camara: Causa de una doenca das folhas algumas videiras. Agron. Lusit. 17:297–333.

(Prepared by J. R. McGrew and F. G. Pollack)

Zonate Leaf Spot

Zonate leaf spot, also called target spot, has been reported infrequently on wild and cultivated grapes. The fungus that causes this disease has a wide host range of both herbaceous and deciduous plants and is known to be present from Florida to Massachusetts and north into Canada and in Japan, China, and India. It has been reported to produce severe defoliation on several crop plants.

The disease may be extensive in a vineyard one year yet not appear the following year. All grape cultivars appear equally susceptible. A severe outbreak of zonate leaf spot in a vineyard may originate from any of several species of nearby infected trees or shrubs. Field studies have shown that the fungus can move limited distances (less than 50 m).

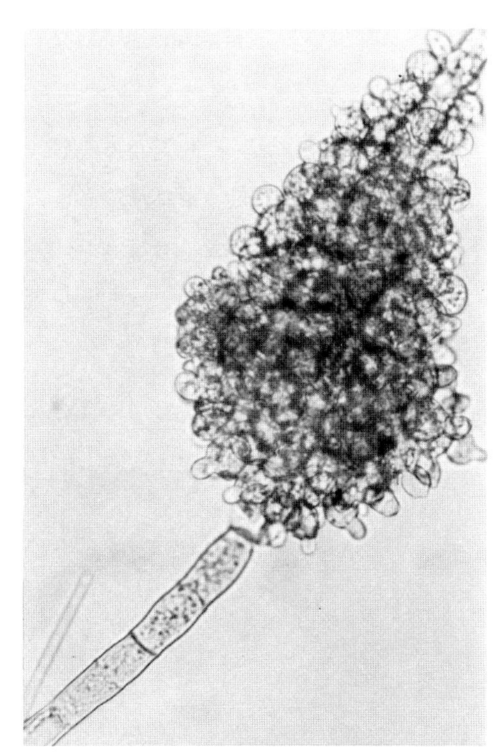

Fig. 24. Fruiting structure of *Cristulariella moricola*. (Courtesy A. J. Latham)

Symptoms

The generally circular lesions of zonate leaf spot may appear at any time during the growing season following several days of high humidity. They resemble lesions of leaf blotch except that the larger lesions tend to have a more concentric zonation (Plate 50). The central portion of older lesions tends to disintegrate, leaving a hole in the leaf. Under some conditions, lesions can be numerous and probably affect vine growth.

Causal Organism

Cristulariella moricola (Hino) Redhead (syn. *C. pyramidalis* Waterman & Marshall) is the causal organism of zonate leaf spot. A sclerotial stage produced by this fungus, known as *Sclerotium cinnamomi* Sawada, may serve as an overwintering stage. The entire fungal fruiting structure, which resembles a miniature conifer, serves as a functional conidium (Fig. 24). It is pale, conical, and up to 0.5 mm long. It is borne on a slender stalk that is only one cell thick and four to many cells long. The stalk produces globose or lobed cells that proliferate by budding, finally forming a dense mass of cells. These cells are not freed individually as conidia.

The entire structure, which can break away from the stalk, may be carried by wind. It can be found affixed horizontally to a leaf surface at the center of a newly formed lesion. The fruiting structures are formed only during periods when the relative humidity is above 96%. Lesions may resume production of these structures even after dry periods of 60 days.

In culture, the conical structures produce minute phialides and globose phialospores (2.4–3.5 μm in diameter) from the globular cells.

Selected References

French, W. J. 1972. *Cristulariella pyramidalis* in Florida: An extension of range and new hosts. Plant Dis. Rep. 56:135–138.

Pollack, F. G., and Waterworth, H. E. 1969. A leafspot disease of Kenaf in Maryland associated with *Cristulariella pyramidalis*. Plant Dis. Rep. 53:810–812.

Redhead, S. A. 1979. Mycological observations: 1. On *Cristulariella*; 2. on *Valdensinia*; 3. on *Neolecta*. Mycologia 71:1248–1253.

Trolinger, J. C., Elliott, E. S., and Young, R. J. 1978. Host range of *Cristulariella pyramidalis*. Plant Dis. Rep. 62:710–714.

(Prepared by J. R. McGrew and F. G. Pollack)

Septoria Leaf Spot

Septoria leaf spot, also called *mêlanose,* is infrequently reported in the eastern United States from New York to Florida and west from Wisconsin to Texas. Presumably introduced into Europe during the phylloxera crisis, it has now been reported from France, Germany, Spain, Switzerland, the European Soviet Union, and Algeria. The disease is known to affect American *Vitis* spp., muscadine grapes, and some *V. labrusca* cultivars. *V. vinifera* cultivars are reported to be immune.

Symptoms

Few to many reddish brown to black spots (Plate 51) appear generally after midseason. They are interveinal, angular, and generally 1–2 mm in diameter, or up to 2 cm in diameter on muscadine foliage. The margins of lesions are often thickened. Later, if there are several spots on a leaf, the surrounding area may become yellow.

Causal Organism

The Septoria leaf spot pathogen is *Septoria ampelina* Berk. & Curt. Pycnidia are 50–60 μm in diameter and have a large, open ostiole. Conidia (40–60 × 2 μm) are hyaline and undulating, with three to six septa.

Selected Reference

Boubals, D. 1983. Une autre maladie de la vigne sevit dans le Penedes (Espagne). Prog. Agric. Vitic. 100:453.

(Prepared by J. R. McGrew and F. G. Pollack)

Other Minor Foliage Diseases

Brulure

Caused by the fungus *Anthostomella pullulans* (de Bary) Bennett, *brulure* occurs on all parts of the vine. The causal fungus is distributed worldwide and is generally a saprophyte.

Cladosporium Leaf Spot

Cladosporium viticola Cesati causes a leaf spot disease on older leaves of both wild and cultivated vines. The disease has been reported in Europe and in the eastern United States.

Cercospora Leaf Spot

Cercospora leaf spot (*cercosporiose*), caused by *Phaeoramularia dissiliens* (Duby) Deighton, produces variable yellowish to dark spots. It has been reported from Pakistan through Asia Minor, Europe, and North Africa.

Tar Spot

Tar spot, caused by *Rhytisma vitis* Schw., produces a black spot 2–4 mm in diameter, with or without a circular brown halo up to 1 cm in diameter. It is common on wild vines in the southeastern United States.

Miscellaneous Leaf Spots

A leaf spot caused by *Asperisporium minutulum* (Sacc.) Deighton produces lesions with indefinite margins. It occurs only on *V. californica* in California and Oregon.

Another leaf spot, caused by *Phaeoramularia heterospora* (Ell. & Gall.) Deighton, has been reported on *V. californica* and *V. girdiana* in California and on *V. vinifera* in Israel.

Rupestris Speckle

Rupestris speckle appears to be a physiological disorder associated only with *V. rupestris.* It may appear in a milder form on hybrids derived from this species, such as Villard blanc and Chambourcin.

Symptoms are similar to those of Septoria leaf spot or oxidant stipple but tend to be on older and shaded leaves. The necrotic areas vary from circular to angular, are generally less than 2 mm in diameter, and tend to show a yellow halo (Plate 52). Spotting is more pronounced on weak or overcropped vines and may be the result rather than the cause of the stress. No control measures have been prescribed, and the effect on vine growth appears to be insignificant.

Selected References

Bennett, F. T. 1928. On *Dematium pullulans* De B. and its ascigerous stage. Ann. Appl. Biol. 15:371–390.

Deighton, F. C. 1976. Three fungi on leaves of *Vitis*. Trans. Br. Mycol. Soc. 67:223–232.

Griffiths, D. A. 1974. The origin, structure and function of chlamydospores in fungi. Nova Hedwigia 25:503–547.

Viala, P. 1893. Les Maladies de la Vigne. 3rd ed. Coulet, Montpellier, France. 595 pp.

(Prepared by J. R. McGrew and F. G. Pollack)

Wood and Root Diseases Caused by Fungi

Eutypa Dieback

Eutypa dieback, known also as "dying arm" and formerly as "dead arm," is one of the most destructive diseases of the woody tissues of commercially grown grapes. The known distribution of the disease coincides with that of the grapevine throughout most countries of both hemispheres; its frequency in any region is limited only by the incidence of rainfall. In general, the disease occurs abundantly where mean annual rainfall exceeds 600 mm but is unlikely to be found where annual rainfall is below 250 mm. It is equally prevalent in regions where winters are severe, such as central Europe and the eastern United States, and in more temperate regions, such as coastal California, southeastern Australia, southern France, and the Cape Province of South Africa.

The causal fungus has a wide host range including approximately 80 species distributed in at least 27 botanical families. Most of its hosts are tree species, including some that are components of natural forests. The most severely affected horticultural hosts are grapevine, apricot, and black currant.

Symptoms

Eutypa dieback is seldom seen in grapevines less than eight years old, and in areas where incidence of the disease is high, diseased vines become more numerous each year thereafter. The most readily recognized symptoms, most evident during the first two months of the annual growth cycle and especially when the new season's shoots are 25–50 cm long, are deformation and discoloration of the shoots. The young leaves are smaller than normal, cupped, and chlorotic; they often develop small necrotic spots and tattered margins, sometimes with larger areas of necrosis, as they age. A marked dwarfing of the internodes (Plate 53) accompanies the development of these leaf symptoms. Clusters on affected shoots may have a mixture of large and small berries.

The symptoms are readily seen until late spring, when affected shoots become obscured from view by adjacent healthy growth. Nevertheless, symptoms on foliage of diseased arms become more extensive in each successive year until, finally, part or all of the arm fails to produce shoots in spring.

The pathogen does not normally enter the green shoots of the current season's growth, and therefore it cannot be cultured from these tissues. The foliar symptoms are believed to be induced by translocation of a toxin generated in the older wood invaded by the mycelium.

Close examination of an arm, cordon, or trunk with vascular connection to shoots bearing foliar symptoms usually reveals a canker surrounding a pruning wound made several years previously. Removal of the loose bark is necessary to show the extent of the canker (Plate 54). In cross section, a wedge-shaped zone of necrotic sapwood may be found extending from the point of origin of the canker (Plate 55). The dead wood is brown, hard, and brittle.

Causal Organism

Eutypa lata (Pers.:Fr.) Tul. & C. Tul. (syn. *E. armeniacae* Hansf. & Carter, anamorph *Libertella blepharis* A. L. Smith [syn. *Cytosporina* sp.]) is the causal organism. It produces perithecial stromata (Fig. 25) on diseased grapevine wood, at first in small patches surrounding the original site of infection, or sometimes on the wound stub that formed the point of entry, several years after the initial infection. Later, as the vine is more extensively invaded, larger areas of stromatic tissue may form on the surface of the dead wood after the loose bark has fallen away. Infected vine wood that has been allowed to remain on the soil is an especially favorable substrate for development of stromata. The stromata are black and continuous, and perithecia are revealed when a shallow slice is cut from the surface with a sharp blade (Plate 56).

Asci are borne on pedicels (60–130 μm long) and measure 30–60 $\times$ 5–7.5 μm, with an apical pore (Fig. 25). The eight ascospores are pale yellow and allantoid and measure 6.5–11 $\times$ 1.8–2 μm.

E. lata may be readily cultured on common laboratory media from small chips taken aseptically from the margin of discolored sapwood in diseased arms or trunks. White mycelium grows from infected wood chips after three to four days at 20–25° C. Perithecia are not produced in culture, but after six to eight weeks, conidiomata may develop, often exuding the characteristic single-celled conidia (18–45 $\times$ 0.8–1.5 μm) (Fig. 26) in orange cirri. Exposing the culture plates to a 12-hr light-dark regime or to near-ultraviolet radiation promotes sporulation.

Not all isolates sporulate, and isolates vary considerably in the amount of dark pigment produced in the medium after one to two weeks. For these reasons, preliminary diagnosis is most readily accomplished by comparing the gross morphology of colonies five or six days old with that of reliable reference cultures transferred at the same time.

The anamorph may be found on the inner bark covering infected wood. Orange cirri containing conidia may ooze from

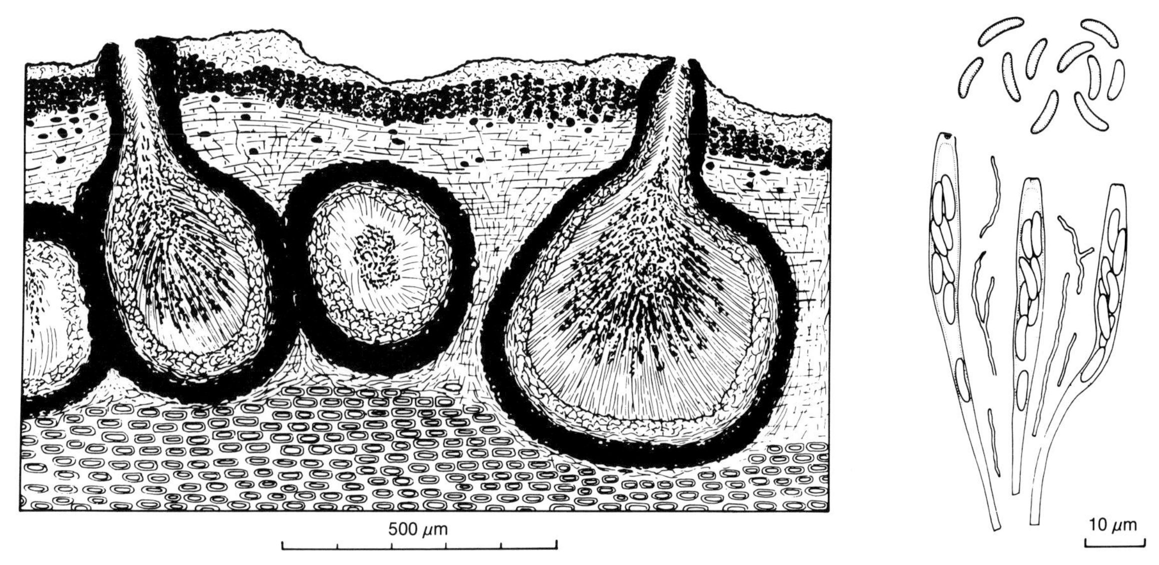

Fig. 25. Vertical section of perithecial stroma (left) and asci and ascospores (right) of *Eutypa lata*. (Reprinted, by permission, from Carter and Talbot, 1974)

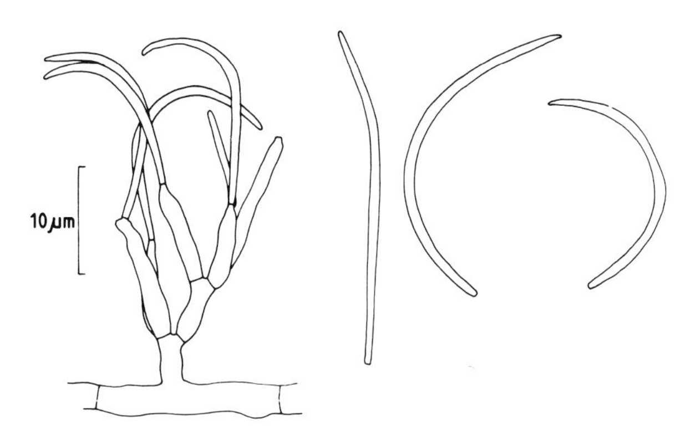

Fig. 26. Conidiogenous cells and conidiophores (left) and conidia (right) from a culture of *Libertella blepharis*. (Drawing by B. C. Sutton)

this tissue following incubation in a moist chamber. The spores of the anamorph do not normally germinate on laboratory media, and there is no evidence that they function as propagules. It is possible that they function as spermatia.

Disease Cycle and Epidemiology

In regions where winters are temperate, perithecia of *E. lata* reach maturity early in spring, and ascospores are disseminated with each rainfall of more than about 1 mm. By late autumn the perithecia are almost exhausted, but nevertheless, sufficient ascospores are available to infect vines pruned during the following winter. In regions where temperatures below 0°C prevail in winter, dissemination of ascospores is greatest in late winter, and they are therefore in abundant supply at the time when grapevines are usually pruned. Studies in the Central Valley of California suggest that viable ascospores may travel up to 50–100 km.

Infections are initiated when ascospores enter freshly made wounds. Rain is a requisite for the release of ascospores and, after aerial transport and deposition, for their entry into the open ends of vessels exposed by pruning. The susceptibility of wounds diminishes markedly during the two weeks following pruning, and after four weeks the wounds are unlikely to be infected.

Ascospores germinate in 11–12 hr at the optimal temperature of 20–25°C. Germination occurs within the vessels, usually 2 mm or more beneath the wound surface. The mycelia proliferate slowly, at first within the vessels and later through associated elements of the functional wood.

The disease develops slowly on grapes, and no symptoms are seen during the first one or two growing seasons after infection. By the third or fourth season, a canker is usually apparent, often accompanied by the foliage manifestations previously described. Several more years may elapse before the affected arm or trunk is killed. Because of the slow progress of the disease, its full economic impact is not likely to be felt until a vineyard reaches maturity. Perhaps the greatest threat to vine productivity posed by Eutypa dieback is the possibility of infection of the many large wounds made when mature vines are reworked to change the cultivar or to adapt the growth pattern for mechanical harvesting.

Control

In regions where inoculum is produced abundantly on many alternative hosts, it is impossible to manage Eutypa dieback effectively by sanitation methods alone. However, in regions with vast plantings of grapes and few alternative hosts, sanitation may be beneficial. Unfortunately, the cultural requirement for regular pruning provides a multitude of entry points for the pathogen each year. No grape cultivars are known to be immune, but where differences in cultivar tolerance are known, it is advisable to prune the least tolerant cultivars when inoculum levels are low.

Because of the high incidence of the disease in the eastern United States, many growers have adopted a multiple trunk-training system or are practicing a program of trunk renewal from latent buds every 10–15 years.

None of the chemicals used routinely to control other fungal diseases of the grapevine provides protection against *E. lata*, nor indeed is the timing of their application suitable for preventing infection. Furthermore, the slow growth of the pathogen and the delayed manifestation of symptoms make recognition of the disease difficult until extensive invasion has occurred, by which time it is usually too late for effective remedial surgery. Hence, the disease has remained essentially uncontrolled.

Fortunately, the fungicide benomyl provides a highly effective barrier against the invasion of pruning wounds by mycelia from germinating ascospores if sufficient chemical is present in the tissue below the pruning wounds before the spores arrive. To accomplish this, each wound must be flooded to saturation to ensure that the chemical is carried well into the exposed vessels at the wound surfaces; a sparse application of spray fluid cannot be compensated for by increasing the concentration of the chemical in the spray mixture.

Because of these exacting conditions, applications of

benomyl by conventional spraying machines have not been successful. Manual treatment of individual wounds at the time of pruning, or the use of spraying secateurs, which facilitate the treatment of any selected wound with a saturating deposit of spray fluid, are effective means of application.

Because *E. lata* seldom invades annual wood, treatment of the wounds on spurs or stubs left from pruning seasonal growth may safely be omitted, but it is essential to treat all wounds in wood two years of age or older, especially the large wounds made in trunks during renovation or change of cultivar.

Selected References

Bolay, A., and Carter, M. V. 1985. Newly recorded hosts of *Eutypa lata* (=*E. armeniacae*) in Australia. Plant Prot. Q. 1:10–12.

Bolay, A., and Moller, W. J. 1977. *Eutypa armeniacae* Hansf. & Carter, agent d'un grave dépérissement de vignes en production. Rev. Suisse Vitic. Arboric. Hortic. 9:241–251.

Carter, M. V., and Perrin, E. 1985. A pneumatic-powered spraying secateur for use in commercial orchards and vineyards. Aust. J. Exp. Agric. Anim. Husb. 25:939–942.

Carter, M. V., and Talbot, P. H. B. 1974. *Eutypa armeniacae.* Descriptions of Pathogenic Fungi and Bacteria, No. 436. Commonwealth Mycological Institute, Kew, Surrey, England.

Carter, M. V., Bolay, A., and Rappaz, F. 1983. An annotated host list and bibliography of *Eutypa armeniacae.* Rev. Plant Pathol. 62:251–258.

Carter, M. V., Bolay, A., English, H., and Rumbos, I. 1985. Variation in the pathogenicity of *Eutypa lata* (=*E. armeniacae*). Aust. J. Bot. 33:361–366.

Moller, W. J., and Kasimatis, A. N. 1980. Protection of grapevine pruning wounds from Eutypa dieback. Plant Dis. 64:278–280.

Moller, W. J., and Kasimatis, A. N. 1981. Eutypa dieback of grapevines. Pages 57–61 in: Grape Pest Management. D. L. Flaherty, F. L. Jensen, A. N. Kasimatis, H. Kido, and W. J. Moller, eds. Publ. 4105. Division of Agricultural Sciences, University of California, Berkeley. 312 pp.

Pearson, R. C. 1980. Discharge of ascospores of *Eutypa armeniacae* in New York. Plant Dis. 64:171–174.

Ramos, D. E., Moller, W. J., and English, H. 1975. Production and dispersal of ascospores of *Eutypa armeniacae* in California. Phytopathology 65:1364–1371.

Rappaz, F. 1984. Les espèces sanctionnées du genre *Eutypa* (Diatrypaceae:Ascomycetes) étude taxonomique et nomenclaturale. Mycotaxon 20:567–586.

(Prepared by M. V. Carter)

Esca and Black Measles

Esca was one of the earliest described diseases of grapevine. It was noted around the Mediterranean in Roman times. It results in a withering of the plant, presumably because of degradation in the wood.

The pathogenic role of the wood rot organisms to which the disease is usually attributed is still unclear. Furthermore, ambiguity remains between esca reported in Europe and the black measles disease of California. Although they have slightly different symptoms, which are probably the result of cultural and varietal differences, these two diseases may actually be the same.

The disease is most frequent in warm, temperate zones. Because of changes in cultural techniques and the gradual abandonment of sodium arsenite treatment, esca is increasing in Europe.

Symptoms

Symptoms may appear on all or part of the vine. There are two forms of the disease: one chronic, characterized by foliage deterioration, the other acute, characterized by sudden death (apoplexy) of the vine.

Foliage deterioration is the most frequent manifestation. Symptoms appear after bloom, during summer or early autumn, beginning on leaves at the base of shoots and then spreading to all leaves. The leaves have yellowish (white cultivars) or reddish (black cultivars) patches. As the necrotic centers of these patches coalesce, large zones of deterioration form between the veins and the margin of the leaf (Plate 57). The leaves may dry gradually and fall prematurely.

Symptoms on berries differ depending on region and cultivar. In France and northern Italy, the clusters of grapes appear normal, but berries do not fill properly and generally do not reach maturity. In California, southern Italy, and Switzerland, affected berries have brownish violet patches distributed in the epidermis (Plate 58). These berries may remain turgid until maturity or may crack and dry. Symptoms on berries may be present without symptoms on leaves and vice versa. Symptoms on herbaceous organs do not necessarily occur each year.

The most spectacular aspect of the disease in Europe is the sudden death of all or part of the vine, usually observed during very hot periods (Plate 59). The foliage and clusters of grapes dry suddenly in a few days. Desiccation usually begins at the extremity of the shoots. Sudden defoliation of some or all of a vine also occurs in California, but the vine does not die; instead, it usually develops new foliage. This symptom in California is more common in May or June than in the hottest part of summer.

In the wood of affected trunks and arms, a characteristic zone of necrosis generally is associated with a large wound. A cross-sectional cut reveals a central, damaged zone, which is light in color and soft in texture, surrounded by an area of darker, harder wood (Plate 60). A longitudinal section reveals a zone of light, necrotic wood, which is usually preceded by an area of hard, black wood. Sometimes the necrotic zone is sectorial because of secondary infection by fungi associated with esca of wood already invaded by *Eutypa lata,* the causal agent of Eutypa dieback. This combination of symptoms in the wood causes confusion between these two diseases.

Causal Organism

Although esca has been known for a long time, the causal organism(s) is unknown. *Stereum hirsutum* (Willd.) Pers. and *Phellinus igniarius* (L.:Fr.) Quél. are the fungi most regularly associated with the disease, but there is no agreement as to their pathogenic role. Recent work in France shows that these two fungi are frequently isolated from zones of deteriorated wood that is light in color and soft in texture. However, depending on the geographic region, one fungus may predominate. In Bordeaux, as in California and Italy, *Phellinus* is usually predominant, whereas in southern France, *Stereum* is usually isolated. Other fungi, such as *Cephalosporium,* are easily obtained from the hard, brown zones of diseased wood by plating tissue on common nutritive media. Fructification of *Stereum* and *Phellinus* in vineyards has been observed only on dead wood. However, inoculation of cuttings in the greenhouse or of vines in the vineyard using these fungi has not successfully reproduced the symptoms of the disease; thus, their role in the disease remains uncertain.

Disease Cycle and Epidemiology

Until the causal organism(s) is identified, a disease cycle and epidemiologic data cannot be established.

Control

Large pruning wounds should be avoided because they provide entry points for potential pathogens. Burning of dead vines is advisable. Replacing the diseased trunk using a renewal shoot from the base of the vine has given excellent results.

Where permitted, chemical control using a sodium arsenite drench is very effective. A single treatment is applied to dry vines during dormancy, at least 10 days after pruning and two to three weeks before bud burst to prevent phytotoxicity. All vines in the vineyard where symptoms of the disease have been found should be treated. Treatment should be repeated for two or three years. Treatment is also recommended on vines where pruning has been severe (rejuvenation, or retraining of vines for mechanical harvesting).

An alternative to sodium arsenite, where allowed, is DNOC (4,6-dinitro-*o*-cresol). Two treatments are advisable, the first after pruning and the second during sap flow in spring.

Selected References

Baldacci, E., Belli, G., and Fogliani, G. 1962. Osservazioni sulla sintomatologia e sull'epidemiologia della carie del legno di vite (mal dell'esca) da *Phellinus igniarius* (L. ex Fr.) Patouillard. Riv. Patol. Veg. (Ser. 3) 2:165–184.

Bisiach, M., and Vercesi, A. 1984. Problemi connessi con le malatie del legno della vite causate da funghi. Atti Accad. Ital. Vite Vino Siena 36:113–122.

Chiarappa, L. 1959. Wood decay of the grapevine and its relationship with black measles disease. Phytopathology 49:510–519.

Dubos, B., Roudet, J., and Dumartin, P. 1985. Mise au point d'actualité sur les maladies de dépérissement de la vigne. Pages 301–309 in: Premières Journées d'Études sur les Maladies des Plantes. Association Nationale pour la Protection des Plantes, Versailles. 412 pp.

Geoffrion, R. 1971. L'esca de la vigne dans les vignobles de l'ouest. Phytoma 23(366):21–31.

Viala, P. 1926. Recherches sur les maladies de la vigne: Esca. Ann. Epiphyt. 12:5–108.

(Prepared by B. Dubos and P. Larignon)

Black Dead Arm

Black dead arm disease has been observed in Hungary, especially in the Tokaj region, since 1974. It also occurs near Naples, Italy, and the causal fungus has been found in wood of *V. labrusca* 'Concord' vines in Canada. Increase in the disease has been associated with changes in training systems, such as a shift from freestanding vines to trellised vines.

Symptoms

A mild chlorosis may appear on leaves, depending on the rate of vascular involvement. If water transport becomes insufficient during the growing season, leaves wilt.

In general, neither the clusters nor the berries are infected during the growing season in Tokaj. However, the fungus has been reported to cause a severe berry and cluster rot on *V. vinifera* cultivars White Hanepoot and Red Hanepoot in South Africa. Berries there are infected near ripening and become dark brown, shriveled, and mummified. Yield losses of 25–30% have been reported.

Narrow black streaks develop in the xylem of infected spurs, arms, and trunks, but rarely in one-year-old canes. The streaks, initially 3–5 mm wide, expand toward the pith as well as longitudinally and laterally. Longitudinal expansion is most rapid. The black tissue becomes nonfunctional and necrotic. The bark over the diseased xylem also dies. In cross section, the affected tissue shows a characteristic sectorial black discoloration, sometimes reaching the pith. The pathogen grows in the vascular elements as well as in the adjacent cells. In natural splits of the outer bark, numerous black pycnidia develop singly or in groups. Affected vines fail to break dormancy or wilt suddenly during the growing season.

The disease has also been observed on the scion wood of bench grafts of Red Traminer/5BB (Plate 61) after winter storage. Affected parts may be covered with pycnidia. Although tissue above the graft union generally dies, the rootstock may remain healthy.

Causal Organism

Botryosphaeria stevensii Shoem. (syn. *Physalospora mutila* (Fries) N. E. Stev., anamorph *Sphaeropsis malorum* Berk. [syn. *Diplodia mutila* (Fries) Mont.] causes black dead arm. Hyphae, initially hyaline but later brown, are septate. Pycnidia (130–195 μm in diameter) develop singly or sometimes in groups in dead cortical tissue. They have long beaks (33–195 μm long) with ostioles, and their walls are dark brown and rather thick.

Conidia ($24.0–27.3 \times 10.1–13.0$ μm) are hyaline, one-celled, and cylindric, with thick, smooth, glassy walls and broadly rounded ends. Guttules are present in the cytoplasm. Conidia remain hyaline for more than 15 days after discharge under humid conditions and en masse appear white, sparkling, and granulated around the ostioles. Occasionally conidia are exuded in short tendrils and slowly discolor to light brown. Two-celled, brown conidia occur rarely.

Perithecia of the pathogen have not been found on grapevines in Hungary.

Disease Cycle and Epidemiology

B. stevensii overwinters in diseased woody parts of vines, and pycnidia develop in spring and autumn during rainy periods. The pathogen may invade tissue through mechanical injuries such as pruning wounds. Penetration through pruning wounds is aided by bleeding in spring because the plant sap keeps the wounds wet for a longer period of time.

Infection occurs in the temperature range of 15–26° C but is optimal between 23 and 26° C.

Control

Information is lacking on chemical control of black dead arm. Removal and destruction of diseased vine parts are recommended.

Selected References

Chamberlain, G. C., Willison, R. S., Townshend, J. L., and De Ronde, J. H. 1964. Two fungi associated with the dead arm disease of grapes. Can. J. Bot. 42:351–355.

Cristinzio, G. 1978. Gravi attacchi di *Botryosphaeria obtusa* su vite provincia di Isernia. Inf. Fitopatol. 28:23–25.

Lehoczky, J. 1974a. Black dead arm disease of grapevine caused by *Botryosphaeria stevensii* infection. Acta Phytopathol. Acad. Sci. Hung. 9:319–327.

Lehoczky, J. 1974b. Necrosis of nurseried grapevine grafts of *Botryosphaeria stevensii* infection. Acta Phytopathol. Acad. Sci. Hung. 9:329–331.

Shoemaker, R. A. 1964. Conidial states of some *Botryosphaeria* species on *Vitis* and *Quercus*. Can. J. Bot. 42:1297–1301.

Verwoerd, L., and Dippenaar, B. J. 1930. On the occurrence of a berry wilt and rot of grapes (*Vitis vinifera*) caused by *Sphaeropsis malorum* Berk. S. Afr. Dep. Agric. Sci. Bull. 86:1–16.

(Prepared by J. Lehoczky)

Armillaria Root Rot

Armillaria root rot is an important disease found throughout the temperate areas. The causal fungus has been reported on over 500 species of plants from 82 countries. The fungus is called the mushroom fungus, the shoestring fungus, the honey fungus, the oak root fungus, and *der Hallimasch*. The disease it causes is called mushroom root rot, shoestring root rot, and at times when referring to grapes, *pourridié*. The last name suggests that the disease might be confused with that resulting from infection by *Dematophora necatrix,* which is also called *pourridié* in France, and it appears that in early reports, this was true.

Although the disease is apparently a serious problem in France, it is less serious in most other grape-growing areas. In California before the use of methyl bromide, it was a problem where grapes were planted on land that had previously been planted with fruit trees.

Symptoms

Infected grapevines may die quickly and in the process show

severe wilting. Infection may also result in a slow decline accompanied by lack of vigor; stunting; small, dark green foliage; and eventual death of the plant. Stunted plants may live through a season but frequently die during the dormant period. Leaves may lose their green color or wilt, and if so, they may show typical symptoms of sunburn. Commonly, a number of plants show varying degrees of decline in localized areas in a vineyard.

Positive identification of the fungus is made by scraping away the bark on the trunk, at or below the soil line, or on the larger roots. The fungus produces white mats between the bark and hardwood (Plate 62). The signs on grape trunks are not typical of the fungus in that the bark pattern on grape causes the mats to form in striations rather than as a solid mat or plaque, as is characteristic on other hosts. In the roots, however, the typical layers of white fungus tissue are formed between the bark and the wood.

Infected tissues have a distinct mushroomlike odor when moist. Rhizomorphs—black fungus strands that look somewhat like roots—may be formed on the outsides of roots (Plate 63). Under California conditions, they are only found attached to the roots and do not grow into the soil.

During the cool part of fall or early winter, the fungus may produce mushroom fruiting bodies (Plate 64) at the soil line around the trunk of infected vines; rarely, the mushrooms may be found attached to a root that is near the soil surface. The mushrooms if present are helpful in diagnosing the disease but frequently are not produced.

Causal Organism

The causal fungus, *Armillaria mellea* (Vahl:Fr.) Kummer (syns. *Agaricus melleus* Vahl, *Armillariella mellea* (Vahl:Fr.) Karst.), is distinguished by its mushroom fruiting bodies. These can vary from 4 to 28 cm in diameter, depending on the number of fruiting bodies produced in a clump; the more that are produced, the smaller they are. They also vary in color; though usually honey-colored, they may be lighter or considerably darker. Sometimes there are darker colored scales on the top of the mushroom cap. The mushrooms have an annulus (a ring of tissue where the cap was attached to the stalk before it expanded), which also varies in size.

The fungus also is distinguished by the production of true rhizomorphs. If fruiting bodies and rhizomorphs are not formed, the fungus can be identified by the large white plaques of hyphae found beneath the bark at or below the soil line.

Disease Cycle and Epidemiology

Although *A. mellea* infects roots, it is not typically soilborne because it exists only in woody plant materials in the soil. Susceptible roots that contact this colonized food base are penetrated by rhizomorphs, principally by mechanical pressure. The fungus moves from one plant to the next by root contact. In grapes, depending on spacing, it tends to move down the rows, although in older vineyards it will move from row to row. If only roots are infected, damage is slight, but the fungus may eventually move up the roots to the trunk, where it girdles the plant and kills it. Spores produced by mushrooms rarely if ever are responsible for spread of the fungus.

Movement of broken infected roots by equipment or moving water is another way the fungus is spread. In California orchards and vineyards, old stream beds can sometimes be detected by following the pattern of disease.

The disease occurs on plants in a number of different soil types, but in California it tends to be more severe in heavy soils. The fungus is able to survive in most soils that support the host.

Control

There are currently no systems for forecasting Armillaria root rot. However, knowledge of the history of a given soil sometimes is helpful in assessing the risk to new vineyards. Examination of roots of previously grown plants during preparation of land for planting can determine the presence of the pathogen. If a previous crop is known to have had the fungus, control by fumigation may be necessary, for *A. mellea* can live a long time in old roots. Because the fungus persists in dead roots that may be deep in the soil and that become brittle as they decay, removal of these roots is difficult and treatment with chemicals is less effective.

Two fumigants, carbon bisulfide and methyl bromide, are used; however, methyl bromide has been shown to be more effective. In some soils, deep placement (60 cm) is necessary, which makes application more difficult and costly. At less than lethal dosages of these fumigants, *A. mellea* is sufficiently weakened to be attacked by *Trichoderma* spp. *Trichoderma* alone, when added to soil, has been reported to reduce the amount of infection.

Although little work has been done on resistant rootstocks, the discovery of resistance in *Vitis* would offer a valuable means of control.

Selected References

Ohr, H. D., Munnecke, D. E., and Bricker, J. L. 1973. The interaction of *Armillaria mellea* and *Trichoderma* spp. as modified by methyl bromide. Phytopathology 63:965–973.

Pegler, D. N., and Gibson, I. A. S. 1972. *Armillariella mellea.* Descriptions of Pathogenic Fungi and Bacteria, No. 321. Commonwealth Mycological Institute, Kew, Surrey, England.

Raabe, R. D. 1962. Host list of the root rot fungus *Armillaria mellea.* Hilgardia 33:25–88.

Raabe, R. D. 1979. Testing grape rootstocks for resistance to the oak root fungus. Calif. Plant Pathol. 46:3–4.

Thomas, H. E. 1934. Studies on *Armillaria mellea* (Vahl) Quel., infection, parasitism and host resistance. J. Agric. Res. 48:187–218.

Watling, R., Kile, G. A., and Gregory, N. M. 1982. The genus *Armillaria*—Nomenclature, typification, the identity of *Armillaria mellea* and species differentiation. Trans. Br. Mycol. Soc. 78:271–285.

(Prepared by R. D. Raabe)

Phymatotrichum Root Rot

Phymatotrichum root rot, commonly called Texas root rot, occurs on grapes and many species of cultivated dicotyledonous plants. Monocotyledonous plants and winter-grown annual crops are not affected by this disease. The disease is known to occur from the southwestern United States (Texas, New Mexico, Arizona, southern Nevada, and southeastern California) to northern and central Mexico, where it causes economic losses in crops such as alfalfa, apples, cotton, peaches, and pecans. The pathogen is indigenous to the semidesert areas of the United States and Mexico but has recently been reported in mango and avocado orchards in the semitropical states of Mexico (Veracruz, Michoacan, and Sinaloa).

Symptoms

Phymatotrichum root rot usually appears in a circular pattern in the vineyard. Affected vines may suddenly wilt and die during early summer or midsummer. Leaves of vines rapidly killed by the disease become brown and brittle and remain rigidly attached to the dead plant (Plate 65). Before wilting, leaves on some infected vines may gradually turn yellow or red, with interveinal yellow spots and irregular necrotic areas on the leaf blade and margins. Some diseased vines may show discoloration and necrosis of a few leaves in early summer, then no further symptoms for the rest of the season. Shoot growth is reduced and leaves are dull green on such vines during the following year.

Vines showing complete foliar discoloration or necrosis in early summer become partly or totally defoliated by midsummer, and their clusters become exposed and sunburned. Occasionally, these vines show signs of apparent recovery during late summer and produce new shoots. Such plants may die during the winter or survive but decline the following year.

Many grape roots have been rotted by the time the first foliar symptoms are observed. Thick or thin mycelial strands of the pathogen are usually observed on the surfaces of infected roots (Plate 66). Often, the root cortex is easily separated from the woody cylinder.

Causal Organism

The causal fungus, *Phymatotrichum omnivorum* (Shear) Duggar (syn. *Ozonium omnivorum* Shear), produces large, sparingly branched, leader hyphal cells (Fig. 27). White mycelial strands, which become brown with age, are formed by parallel growth of hyphae and lateral branching of short hyphal cells. The fungus can be easily identified under the microscope by observing the strands, which have numerous acicular and characteristically cruciform (cross-shaped) hyphae (Fig. 27).

Globose to irregularly shaped brown sclerotia (1–2 mm in diameter) are produced singly or in clusters in soil from infected roots. Sporemats, which consist of numerous hyphae with globose to elongated conidiophores bearing single-celled, spherical to ovate conidia (4.8–5.5 μm in diameter) (Fig. 27), are sometimes produced on the surface of the soil after several days of irrigation or rain.

Disease Cycle and Epidemiology

Sclerotia are produced in soil following infection. They are the more resistant survival structures of the fungus and are the source of primary inoculum. They are usually produced 15–75 cm deep in the soil profile but can be found at a depth of 2 m. They can survive in soil for 12 years. Sclerotia germinate and produce mycelial strands that colonize root surfaces and form infection cushions at lenticels and cracks where lateral roots emerge.

The fungus infects the root cortex, then invades the vascular elements of the root, impeding water translocation. Sudden wilt of the plant occurs as a result of either extensive killing of the root system or invasion and occlusion of the vascular tissue of the main roots.

Plants seem to be killed first where inoculum of the fungus is most abundant or where fungal growth is unrestricted. The fungus moves from diseased to neighboring healthy plants by colonization of roots that overlap. However, the disease often moves slowly for many years and appears to be restricted to certain areas of the vineyard. The reason for this pattern of disease spread in vineyards and other crops has not been explained.

Optimal temperatures for growth of *P. omnivorum* (28–30° C) occur in soil only during summer months. In cotton, the disease is less severe in soils with a pH of 6 or lower, and it is a serious problem where soil is slightly acid to alkaline (pH 6.5–8.5) and calcareous.

Control

To control Phymatotrichum root rot, avoid planting in infested soil by mapping the areas where plants of a susceptible crop like alfalfa or cotton show symptoms. Use only healthy propagation material when planting in pathogen-free soil. Roots with partially rotted tissue may carry hyphal strands on their surface and mycelium in their vascular tissue and thereby introduce the pathogen.

Replant vacant sites with vines grafted onto Dog Ridge rootstock. This rootstock has good vigor and root regeneration characteristics that appear to enhance its survival compared to *V. vinifera* cultivars on their own roots. However, excess vigor and low fruitfulness may be a problem with some cultivars on this rootstock in fertile soils.

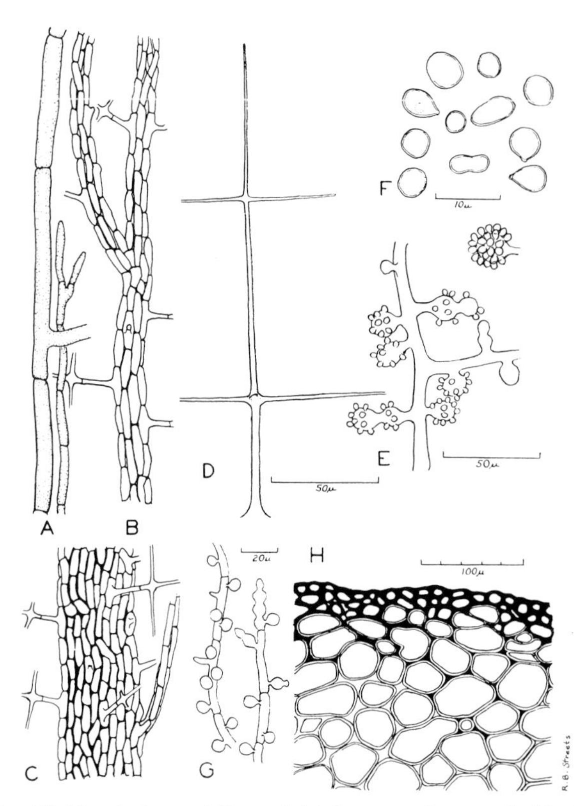

Fig. 27. Morphology of *Phymatotrichum omnivorum*. **A,** Single hypha. **B,** Strand formed by branching hyphae surrounding a large central hypha. **C,** Mature strand with cruciform hyphae. **D,** Cruciform hypha with acicular points. **E,** Conidia on conidiophores from a sporemat. **F,** Conidia freed from conidiophore. **G,** Conidiophores before conidiation. **H,** Cross section of a sclerotium, showing thick-walled rind cells and parenchyma. (Reprinted, by permission, from Streets and Bloss, 1973)

Selected References

Herrera, P. T. 1984. Investigación sobre portainjertos y su resistencia a pudrición de la raíz por *Phymatotrichum omnivorum*. Pages 83–93 in: En Memorias ler. Simposia Internacional sobre Pudrición Texana, Biologia, y Control de *Phymatotrichum omnivorum* (Shear) Duggar. Escuela de Agricultura y Ganadeŕia, Hermosillo, Son. Mêxico-Dic. 99 pp.

Lyda, S. D. 1978. Ecology of *Phymatotrichum omnivorum*. Annu. Rev. Phytopathol. 16:193–209.

Mortensen, E. 1938. Nursery tests with grape rootstock. Proc. Am. Soc. Hortic. Sci. 36:153–157.

Mortensen, E. 1952. Grape rootstock for southwest Texas. Tex. Agric. Exp. Stn. Prog. Rep. 1475. 11 pp.

Perry, R. L. 1980. Anatomy and morphology of *Vitis* roots in relation to pathogenesis caused by *Phymatotrichum omnivorum*. Ph.D. thesis. Texas A&M University, College Station. 205 pp.

Streets, R. B., and Bloss, H. E. 1973. Phymatotrichum root rot. Monograph 8. American Phytopathological Society, St. Paul, MN. 38 pp.

(Prepared by T. Herrera and S. D. Lyda)

Verticillium Wilt

Verticillium wilt occurs sporadically in several countries where grapes are grown. It was first described in the 1950s in Germany on vines growing on American rootstocks. Because affected vines exhibited symptoms similar to those of other diseases and stresses, the disease was misdiagnosed in many other countries where it was actually present. During the rapid expansion of vineyards in California in the 1970s, the disease developed in many of the new vineyards where vines were planted at sites where crop plants susceptible to *Verticillium* had been previously grown.

Symptoms

In the early part of the growing season, infected vines do not show symptoms, but as temperatures rise and soil moisture declines, a few shoots start to die and the vascular elements in the wood of these shoots become discolored (Plate 67). New shoot growth from the base of such vines is often vigorous. Other parts of the affected vine may show no symptoms.

By early summer the leaves on declining shoots wilt and show marginal burning. By midsummer many shoots that had developed normally during the early season collapse completely (Plate 68). The leaves on these shoots become desiccated, and some may drop from the vine. Fruit clusters at the base of these affected shoots dry up, and the individual berries remain attached to the pedicels as shriveled mummies. The degree of collapse varies; some vines show only a few dying shoots, others show collapsed shoots on one side of the vine only, and in a few cases all the shoots on a vine collapse.

Causal Organism

Verticillium dahliae Kleb., the causal fungus, grows rapidly on potato-dextrose agar at 24° C. Mycelium is hyaline to whitish at first but later turns black with the formation of microsclerotia. Microsclerotia arise from individual hyphae by repeated budding. They are dark brown to black and torulose or botryoidal. They consist of swollen, almost globular cells that vary in shape and size (15–50 μm in diameter, maximum 100 μm). Chlamydospores are absent, and dark brown resting mycelium is formed only in association with microsclerotia.

Conidiophores are erect, hyaline, and verticillately branched, with three to four phialides arising at each node. Phialides (16–35 $\times$ 1–2.5 μm) are occasionally secondarily branched. Conidia (2.5–8 $\times$ 1.4–3.2 μm) arise singly at apexes of phialides and are ellipsoidal to subcylindrical and single-celled, rarely one-septate.

V. dahliae is distinguished from the closely related species *V. albo-atrum* Reinke & Berth. mainly by the presence of true microsclerotia and by its ability to grow at 30° C.

In late spring, it is difficult to isolate the fungus from shoots that are starting to show symptoms, but as the season progresses, isolation from leaves, petioles, and discolored wood in the canes becomes easier.

Disease Cycle and Epidemiology

Infection apparently takes place through the roots. Verticillium wilt may appear when vines are planted at locations where previous susceptible crops or weeds have stimulated inoculum buildup in the soil. The distribution of affected vines at such sites is erratic. Studies in Germany showed that the fungus is not spread in grapevines by propagating from infected mother vines.

Vines that are planted in infested soil usually do not show symptoms of the disease the first year. Some vines show symptoms during the second year, and new infections show in subsequent seasons. By the fifth or sixth season, vines that have shown symptoms but have not died recover, and the disease is no longer visible in the vineyard. This apparently spontaneous recovery is one reason it took many years of investigation before anyone showed that *Verticillium* infects grapes.

Full production is delayed in vineyards that show early wilting symptoms of the disease, but after vines have reached an age where they no longer show symptoms, yields do not seem to be adversely affected. Unless young vines are killed by the initial attack, after eight to 10 years it is difficult to find the areas in the vineyard where infections occurred.

Verticillium wilt of grapes is uncommon in the southern San Joaquin Valley of California, even though it appears to be caused by the mild (SS-4) strain of the fungus that affects cotton in that area. For unknown reasons, it is more common in the northern San Joaquin Valley and in the Salinas Valley, where previous crops, such as strawberries, apricots, and tomatoes, have favored the increase of the mild strain of the fungus.

Control

Other than avoiding sites in the few locations where the fungus has been known to kill grapevines, specific control measures for Verticillium wilt do not seem to be warranted. Differences among grape cultivars in susceptibility to the fungus have been found in greenhouse studies in California.

Selected References

Braun, A. J. 1953. Ills of the American bunch grapes. Pages 754–760 in: Plant Diseases, The Yearbook of Agriculture 1953. U.S. Department of Agriculture, Washington, DC. 940 pp.

Canter-Visscher, T. W. 1970. Verticillium wilt of grapevine, a new record in New Zealand. N.Z. J. Agric. Res. 13:359–361.

Hawksworth, D. L., and Talboys, P. W. 1970. *Verticillium dahliae.* Descriptions of Pathogenic Fungi and Bacteria, No. 256. Commonwealth Mycological Institute, Kew, Surrey, England.

Schnathorst, W. C., and Goheen, A. C. 1977. A wilt disease of grapevines (*Vitis vinifera*) in California caused by *Verticillium dahliae.* Plant Dis. Rep. 61:909–913.

Thate, R. 1961. Die Apoplexie der Rebe: eine Verticilliose. Mitt. Biol. Bundesanst. Land Forstwirtsch. Berlin-Dahlem 104:100–103.

(Prepared by W. C. Schnathorst and A. C. Goheen)

Dematophora Root Rot

Dematophora root rot is a serious disease on the roots of many herbaceous and woody plants throughout much of the temperate world. It is most common on deciduous fruit trees and grapes. On grape it is commonly known as *pourridié* and also as *aubernage, bianco, blanc, blanc des racines, blanquet, champignon blanc, malbianco, mal nero, marciume radicale bianco, morbo bianco, pourridié de la vigne, pourriture,* Rosellinia root rot, *Weinstockfaule*, white root rot, and *Wurzelpilze*. The disease has been reported mainly from European countries but also on grape in North and South America, Africa, Australia, New Zealand, India, the Soviet Union, and Japan. The fungus has not been reported from grape-growing areas of California, although grapes planted in infested areas in California have become infected.

Symptoms

Symptoms of Dematophora root rot are not very diagnostic. Infected plants may die very rapidly, may die slowly during one season, or may linger into a second year. The vines that linger frequently bear an excessively large crop the year before they die.

Leaves remain attached to plants that die rapidly. On vines that decline gradually, tendrils and leaves are weak and stunted, wilting is common, and shoots may arise from the base. Dead vines can be pulled easily from the soil because of severe deterioration of the roots. Frequently, vines will break off at the soil line, where the fungus has weakened the wood. The bark below the soil line is darkened and sloughs easily. At the root crown, there may be a black, gummy ooze.

Under moist conditions, the fungus produces abundant hyphae on the surface of infected roots, giving them a white, fluffy appearance (Plate 69). The hyphae tend to grow along the smaller roots and often form flattened strands in soil cavities around the roots. As these fungal strands age, they darken and may take on a tan or brown cast.

The fungus grows rapidly in the infected vine and produces small, white plaques scattered throughout the wood (Plate 70). These plaques differ considerably from those of *Armillaria mellea,* which are confined to the area between the bark and the wood (see Armillaria Root Rot).

Pieces of invaded roots or stems, when put in a moist chamber, are rapidly covered with abundant growth of white hyphae. The fungus also may produce sclerotialike masses on the surface of infected tissues.

Causal Organism

The causal organism, *Rosellinia necatrix* Prill. (anamorph *Dematophora necatrix* Hartig), produces nearly spherical, brown to black perithecia, aggregated and embedded in a brown web of hyphae on the host surface. Perithecia are about 1–2 mm in diameter. Young perithecia have a distinct protrusion around the ostiole that is visible with a dissecting microscope, but ostioles on old perithecia are difficult to find. The perithecial stage takes several years to develop and is rarely found.

Asci are cylindrical (8–12 × 250–380 μm), long-stalked, and unitunicate. The eight ascospores are one-celled, cymbiform, straight or curved, and dark brown. They measure 5–8 × 30–50 μm and have a longitudinal slit running parallel to the long axis of the spore for about one-third of its length.

The conidial stage is composed of brown, rigid synnemata up to 1–5 mm high. The stipe is 40–300 μm thick and is often dichotomously branched toward the apex. Conidia (2.5 × 3–4.5 μm) are produced in large numbers.

Microscopic signs of the pathogen include distinct enlargements of the ends of hyphal cells next to the septa (Fig. 28). This occurs most frequently in older hyphae and is helpful in identifying the fungus.

Disease Cycle and Epidemiology

The fungus grows through soil, using roots it has killed as a food base. It is favored by moisture and organic material in the soil, which can also serve as a food base. It is frequently found in soils high in clay content.

Although the *Dematophora* stage may produce many conidia, most researchers have not been able to germinate the spores. Thus spores appear to be of little importance in the dissemination of the fungus. The fungus is spread with infested soil and by infected nursery stock plants and tends to occur in discrete areas of the vineyard.

The fungus withstands drying and can remain viable in air-dried pieces of wood in the laboratory for several years. It grows optimally at 22–28° C but does not grow at 31° C.

Control

Controlling Dematophora root rot is extremely difficult. Many fumigants, such as allyl bromide, ammonium hydrosulfide, bromopicrin, carbon disulfide, carbon tetrachloride, chloroform, chloropicrin, ethylene dibromide, and formalin pentachlorethane have failed to control the disease in small-scale experiments. In field experiments, methyl bromide was ineffective in California but was effective in Israel. Carbendazim and dazomet also have been reported as giving control.

The use of resistant rootstocks is a logical approach to control. Resistance has been reported in *V. cinerea* and *V. vinifera* 'Carignane' and in Solonis, a complex hybrid. On a field plot started in an infested area in California, some survivors included Iona, Red Malaga, Palomino, Dog Ridge, Salt Creek, and St. George. *V. arizonica, V. flexosa,* and several hybrid selections also survived. Further testing of rootstocks is warranted.

Selected References

Berlese, A. N. 1982. Rapporti tra *Dematophora* e *Rosellinia.* Riv. Patol. Veg. 1:1, 5–17, 33–34.

Hansen, H. N., Thomas, H. E., and Thomas, H. E. 1937. The connection between *Dematophora necatrix* and *Rosellinia necatrix.* Hilgardia 10:561–565.

Khan, A. H. 1949. The root disease caused by *Rosellinia necatrix* (Hart.) Berl. Ph.D. thesis. University of California, Berkeley. 139 pp.

Sivanesan, A., and Holliday, P. 1972. *Rosellinia necatrix.* Descriptions of Pathogenic Fungi and Bacteria, No. 352. Commonwealth Mycological Institute, Kew, Surrey, England.

Sztejnberg, A., Omary, N., and Pinkas, Y. 1983. Control of *Rosellinia necatrix* by deep placement and hot treatment with methyl bromide. Bull. OEPP/EPPO Bull. 13:483–485.

Viala, P. 1891. Monographie du Pourridié. Libraire de l'Académie de Médecine, Paris. C. Coulet, Montpellier, France. 124 pp.

(Prepared by R. D. Raabe)

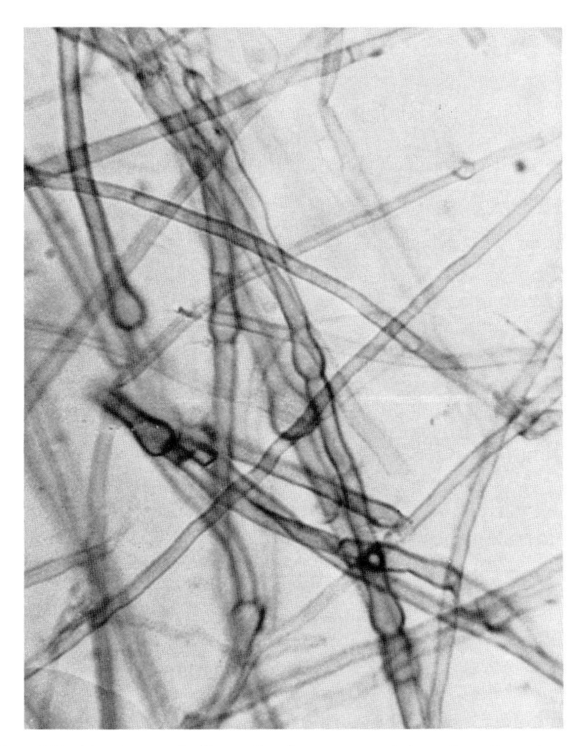

Fig. 28. Hyphae of *Dematophora necatrix,* showing typical swellings at the septa. (Reprinted, by permission, from Compendium of Strawberry Diseases, J. L. Maas, ed., 1984, The American Phytopathological Society, St. Paul, MN)

Phytophthora Crown and Root Rot

Phytophthora crown and root rot occurs throughout the grape-producing regions of the world but is considered a relatively minor disease problem because of its low incidence and sporadic occurrence. It has been observed in South Africa, India, Australia, New Zealand, and the United States (California), most frequently on young vines.

Symptoms

Phytophthora crown and root rot occurs on single vines or small groups of vines, primarily in sections of the vineyard that are poorly drained. It is most common following seasons with abnormally high rainfall or on vines that are excessively irrigated.

Affected vines are smaller than nearby healthy plants and appear stressed. The foliage often becomes chlorotic or colors prematurely in the autumn. A canker develops near the soil line and usually extends downward to the roots but may also extend a short distance upward. Cutting through the canker with a knife reveals necrotic, brown bark that blackens as the tissue decays (Plate 71). Sometimes the decayed bark sloughs off, leaving only the woody vascular cylinder covered by periderm that formed before infection. Removal of the periderm exposes the necrotic woody tissue underneath. Woody roots and fibrous feeder roots may also become infected, blacken, and decay.

Vines with severe root rot or extensive girdling of the trunk may collapse and die. However, if environmental conditions or host response stops lesion expansion before the vine dies, new tissue may be regenerated that allows the plant to recover.

Causal Organism

Several *Phytophthora* species cause crown and root rot. *P. cinnamomi* Rands has been isolated from diseased crown and root tissues in South Africa, India, and Australia and has been shown to be highly virulent. *P. cactorum* (Leb. & Cohn) Schroet., *P. parasitica* Dast., and *P. cryptogea* Pethyb. & Laff. have also been isolated from diseased grapevines in South Africa; however, these species appear to be less virulent than *P. cinnamomi* and are isolated much less frequently. In California, *P. megasperma* Drechs. and an unidentified *Phytophthora* sp. are associated with the disease.

In culture, *P. cinnamomi* produces thick, broad hyphae (8 μm or more in diameter) bearing pronounced swellings or rounded nodules. Thin-walled chlamydospores (averaging 42 μm in diameter) frequently occur in grapelike clusters at the ends of short lateral branches. Sporangia are produced only in nonsterile soil extract solution or in other aqueous solutions under relatively precise conditions. They are nonpapillate and broadly ellipsoidal to ovoid (average 57 μm long × 33 μm wide) and proliferate through the empty sporangium. The species is further characterized by its heterothallic nature (A2 mating types predominate); long (22 × 17 μm), amphigynous antheridia; camellioid or rosette colony patterns on potato-dextrose agar; and cardinal growth temperatures of 5, 24–28, and 32–34° C.

P. megasperma is a diverse species composed of several morphological types. All are characterized by the abundant production of smooth-walled oogonia and predominantly paragynous antheridia in single culture and by the production of ovoid, nonpapillate sporangia that proliferate primarily through the base of the previously evacuated sporangium. Sporangia are not produced on solid media but are produced readily in aqueous solutions. Isolates associated with grapevines belong to a morphological group that has been recovered from a broad range of other hosts. This group is further characterized by the size of the oogonia (generally >40 μm in diameter), a relatively low maximum temperature for growth in culture (30° C), and a radial to rosette colony pattern on cornmeal agar.

Disease Cycle and Epidemiology

The epidemiology of crown and root rot on grapes probably depends somewhat on which *Phytophthora* species is involved. In California infection is assumed to occur from fall through early spring when the ground is wet and soil temperatures are low. Crown rot caused by *P. megasperma* on other woody fruit plants is greatly exacerbated by prolonged periods of saturated or flooded soil, apparently because such conditions favor the production and dissemination of zoospores and decrease host resistance to infection.

Sporangia of *P. megasperma* are formed at soil temperatures of 6–27° C (optimum 12–24° C), but the frequency of indirect germination (zoospore release) is reduced at temperatures above 20° C. Little or no indirect germination occurs above 25° C, and disease does not appear to develop at these higher soil temperatures. It is very difficult to isolate *P. megasperma* from affected grapevines during the summer months in California.

Disease caused by *P. cinnamomi* is also exacerbated by wet soil, although the requirement for prolonged saturation may be less demanding than for infection by *P. megasperma*. It is also probable that optimal temperatures for disease development are higher for *P. cinnamomi* than for *P. megasperma*. Isolates of *P. cinnamomi* from other hosts form sporangia and chlamydospores primarily at temperatures between 20 and 30° C, with little or no production at 15° C or lower. The minimum temperature for chlamydospore germination appears to be 9–12° C, with an optimal range of 18–30° C.

Control

Phytophthora crown and root rot is unlikely to develop if young vines are planted in a site that is not subject to prolonged periods of excessive soil moisture. As vines grow older, they appear to become more resistant to infection; therefore, good water management during the first several years of vine growth usually prevents occurrence of this disease. Where drip irrigation is used, emitters should be located approximately 30 cm away from the vine trunks to reduce the probability of saturating the crown region.

The use of resistant rootstocks is advisable in sites that are infested or likely to be infested with *P. cinnamomi*. Studies in South Africa revealed that Paulsen 1045, P. 1103, and St. George are highly resistant; Richter 110, Ruggeri 140, Metalliko 101-14, and Grezot are moderately resistant; and Salt Creek, Jacques, and several other hybrids are susceptible.

Selected References

Hansen, E. M., Brasier, C. M., Shaw, D. S., and Hamm, P. B. 1986. The taxonomic structure of *Phytophthora megasperma:* Evidence for emerging biological species groups. Trans. Br. Mycol. Soc. 87:557–573.

Marais, P. G. 1979a. Fungi associated with root rot in vineyards in the Western Cape. Phytophylactica 11:65–68.

Marais, P. G. 1979b. Situation des porte-greffes résistants à *Phytophthora cinnamomi*. Bull. Off. Int. Vin 579:357–376.

Moller, W. J. 1981. Phytophthora crown and root rot. Page 81 in: Grape Pest Management. D. L. Flaherty, F. L. Jensen, A. N. Kasimatis, H. Kido, and W. J. Moller, eds. Publ. 4105. Division of Agricultural Sciences, University of California, Berkeley. 312 pp.

Zentmyer, G. A. 1980. *Phytophthora cinnamomi* and the Diseases It Causes. Monograph 10. American Phytopathological Society, St. Paul, MN. 96 pp.

(Prepared by W. F. Wilcox and S. M. Mircetich)

Grape Root Rot

Grape root rot is a problem associated with weakened vines growing in cold, wet soils. The causal fungus was described in the last century in Europe and has since been reported to occur in the eastern United States. It can be a serious problem in replant sites. The fungus is associated with root rot on *Malus*, *Pyrus*, *Cydonia*, *Prunus*, *Salix*, *Tilia*, *Rosa*, and *Paliurus* as well as *Vitis*.

Symptoms

The symptoms of grape root rot are not very diagnostic. Infected vines gradually decline in vigor and eventually die. Signs of the fungus, in particular the distinctive fruiting bodies, are most helpful in diagnosis.

Causal Organism

The grape root rot fungus, *Roesleria subterranea* (Weinmann) Redhead (syn. *R. hypogaea* Thüm. & Pass.), is a cool-temperature fungus, growing best at 10–12° C. On roots apothecia are produced on white to grayish stalks up to 6 mm long. The apothecia (4–4.5 × 1 mm) appear as mouse gray to greenish hemispherical heads (Plate 72) and turn brown to black with age. The apothecia produce abundant filamentous paraphyses and asci containing eight globular to disk-shaped ascospores 5 μm in diameter. The spores are sometimes septate upon germination, producing one or two germ tubes. The asci deliquesce, allowing the spores to mass together in the head, which is at first covered with a peridiumlike mat of hyphae that soon breaks away because of crowding of the freed spores.

Disease Cycle and Epidemiology

R. subterranea is basically an opportunistic saprophyte that colonizes injured or dead roots. From weak or dead roots, the fungus may infect and grow in healthy, living root tissue. In nature, the fungus generally produces mature ascocarps from spring to autumn. The fungus survives in the soil for many years. Spores are distributed vertically and horizontally in the root zone by ground water, soil animals, and cultivation.

Control

In replant sites, particular attention should be given to removing as many remaining roots as possible. However, other than this standard sanitation practice and planting disease-free vines in well-drained soils, little is known about control measures for grape root rot.

Selected References

Arnaud, G., and Arnaud, M. 1931. Pourridié morille (*Roesleria hypogaea* Thüm. et Pass.). Pages 455–465 in: Traité de Pathologie Végétale. 2 vols. Lechevalier et Fils, Paris. 1,831 pp.

Beckwith, A. M. 1924. The life history of the grape rootrot fungus *Roesleria hypogaea* Thüm. & Pass. J. Agric. Res. 27:609–616.

Redhead, S. A. 1984. *Roeslerina* gen. nov. (Caliciales, Caliciaceae), an ally of *Roesleria* and *Coniocybe*. Can. J. Bot. 62:2514–2519.

Viala, P., and Pacottet, P. 1910. Recherches expérimentales sur le *Roesleria* de la vigne. Ann. Inst. Natl. Rech. Agron. Ser. B 9:241–252.

von Thümen, F. 1885. Die Pilze und Pocken auf Wein und Obst. P. Parey, Berlin. Pages 210–212.

(Prepared by W. Gärtel; translated by H. O. Amberg)

Diseases Caused by Bacteria and Bacterialike Organisms

Crown Gall

The bacterial disease crown gall occurs on over 600 species of dicotyledonous plants. One of the earliest reports of this disease was on grapevines in France in 1853, and its infectious nature was first demonstrated by Carvara in Italy in 1897. By the early 1900s, reports from around the world indicated that the disease was a serious problem, particularly on *V. vinifera* cultivars grown in cold climates. Crown gall was first reported in the United States in 1889; however, it was not until 1907 that E. F. Smith showed conclusively that the disease was caused by a bacterium.

Today, crown gall of grape is a serious problem in viticultural regions where *V. vinifera* and interspecific hybrids are grown and where climatic conditions favor freeze injury. Presumably, freeze injury provides the wounds that are necessary for infection. Incidence of disease following freeze injury may range from a few vines to nearly 100% of the vineyard.

In general, *V. labrusca* cultivars have less crown gall than interspecific hybrids or *V. vinifera*. The *V. labrusca* cultivars Niagara, Dutchess, and Isabella are exceptions in that they can become heavily infected. Aurore, Chancellor, and Cayuga White are examples of interspecific hybrids that are often infected. Experience in the United States suggests that all *V. vinifera* cultivars are susceptible.

Symptoms

The major symptom of crown gall is the fleshy galls that are produced in response to infection. Galls are composed of disorganized primary and secondary phloem tissues. Irregularly shaped parenchyma cells and disorganized vascular bundles can also be found in the gall tissue. On grape, galls are mostly found on the lower trunk, near the soil line (Plate 73). However, they may also develop slightly below the soil surface, or aerial galls may extend more than 1 m up into the trellis. A high incidence of gall may occur in nurseries at the basal ends and below-ground disbudded nodes of some cultivars and rootstocks. Galls on lateral roots, however, are uncommon. Large galls may develop rapidly and completely girdle young vines in one season. Small, localized galls or small, pimply galls extending up the trunk are also common (Plate 74). Galled vines frequently produce inferior shoot growth, and portions of the vine above the galls may die.

Gall expression is determined by the extent of the wound, the grape cultivar, and the strain of the pathogen. Current-season galls are first apparent in early summer as white, fleshy callus growths developing near injured areas of the vine. They often develop near the periphery of old galls. The galls turn brown by late summer and in the fall become dry and corky. After one or two years, the dead galls may flake off the vine.

In some cases, a high level of galling has been associated with graft unions (Plate 75). Because it is usually not possible to distinguish galls from abundant callus formation at grafts, the pathogen must be isolated and identified in such cases. Galls at graft unions have been attributed to inoculum from grafting tools but may also come from systemically contaminated propagation tissue.

Recently, tumorigenic and nontumorigenic strains of biovar 3 of the bacterium were shown to incite lesions in grape roots. Biovars 1 and 2 do not cause this reaction. In the field these lesions appear as dark sunken areas on current-season roots that are about 3–4 mm in diameter. The significance of nontumorigenic root infections by biovar 3 has not yet been determined.

Causal Organism

Agrobacterium tumefaciens (E. F. Smith & Townsend) Conn, the cause of crown gall, belongs to the family Rhizobiaceae and is primarily identified by its ability to stimulate gall production when inoculated to host plants. It is a gram-negative, rod-shaped bacterium and may be motile or nonmotile. Colonies growing on nonspecific culture media are white, convex, circular, glistening, and translucent.

The three biovars of the pathogen, of which biovar 3 is predominant on grape, are usually characterized with specific laboratory tests. Nonpathogenic strains of the bacterium, identified as *A. radiobacter*, are frequently isolated in nature. They may be detected as cohabitants with *A. tumefaciens* in soil, in apparently healthy plant tissues, and in galls.

The development of selective culture media has made it possible to isolate *Agrobacterium* spp. from soil and plant tissues. Some of the media are selective or differential for a specific biovar of the pathogen, while others permit growth of more than one biovar. However, because colonies of tumorigenic and nontumorigenic *Agrobacterium* look identical and grow equally well on all of the selective media, pathogenicity tests are necessary to confirm that isolated strains are *A. tumefaciens*.

Indicator plants that are commonly used for greenhouse pathogenicity tests include tomato, sunflower, and tobacco (particularly *Nicotiana glauca*). Although many strains of *A. tumefaciens* produce galls on these plants, some have a limited host range. Some grape strains from Greece, for example, infect only grape, whereas most of the strains isolated from grape from other parts of the world are tumorigenic on several indicator hosts. Grapevine strains vary considerably with regard to host range as determined on indicator hosts and the sizes of galls they induce on sunflower and on grape cultivars and rootstocks.

In recent years a great deal of research on the molecular genetics of *Agrobacterium* has been stimulated by the discovery that when the pathogen infects plants, part of the bacterial DNA is incorporated into the plant genome. The bacterial DNA that is transferred to the plant is located on a plasmid and comprises genes that code for pathogenicity, host range, and several other characteristics. *A. radiobacter* does not possess the genes for pathogenicity. Molecular biologists are particularly interested in using *Agrobacterium* to introduce beneficial genes into plants.

Disease Cycle and Epidemiology

It is frequently postulated that most crown gall infections result from inoculum in soil entering the plant through a wound. Reports have substantiated the existence of *Agrobacterium* in agricultural soils. Results from vineyard soil sampling, however, have consistently shown low to undetectable levels of the grape pathogen except in soil directly beneath diseased vines. This pathogen has not been detected in nonvineyard soils. Therefore, the significance of soil as a source of primary inoculum for grape crown gall is unclear. Soil type and climate may influence the survival of the pathogen in soil.

Another source of inoculum is contaminated planting material. Unless precautions are taken during the mist propagation of grapes, for example, the grapes frequently become infested and infected with crown gall bacteria. Analyses of roots of clonal rootstocks and rooted cultivars from nurseries have also shown *A. tumefaciens* associated with plants that exhibit no apparent infections.

Inoculum survives in vineyards within galls and systemically infested vines. The bacterium is detectable in grapevine sap and in the callus and roots of rooted cuttings. It has been suggested that during the spring, when grapevines bleed, the bacterium is flushed up from the roots to the aboveground portions of the vine. Survival in roots and in the rhizosphere may also be important when diseased vines are removed from a vineyard; roots that remain in the soil may harbor the pathogen and provide inoculum for infection of replanted vines.

The systemic nature of *A. tumefaciens* has been further substantiated in research with several grape cultivars in the United States and Europe. The results clearly demonstrate the potential for spreading the pathogen via propagation material. The relative distribution of *A. tumefaciens* in vines at various times of the year is not clearly understood. Research suggests that high levels of the pathogen may survive in roots during the dormant season and that the pathogen is not evenly distributed within the vine. An understanding of the distribution of the pathogen within vines would be helpful for the development of effective sampling methods to be used in indexing programs.

Control

Crown gall has been successfully controlled on some hosts with a biological treatment and with an eradicant chemical. The biological control organism is strain K84 of *A. radiobacter,* which has been used on some plants as a treatment to protect against invasion of the pathogen through wounds. Strain K84 produces an antibiotic that inhibits some strains of the pathogen but unfortunately is not active against the biovar 3 strains that predominate on grape.

Eradicant chemicals used to treat established galls have given variable results. In areas where freeze injury to vines occurs only occasionally, chemicals may be useful for treating galls very early in their development to minimize further development. Eradicants such as kerosene are effective in killing gall tissues, but new galls frequently develop at treated sites the following year.

Because the development of crown gall is closely correlated with the occurrence of freeze injury, management practices that reduce injury are useful in managing the disease. In some areas, growers bury young vines in the fall to reduce freeze injury. "Hilling" of trunks with soil in the fall is practiced as a means of protecting crown tissues from cold temperatures, although viticulturists have questioned the effectiveness of this treatment for reducing injury. Hilling above the union on grafted vines protects buds from freezing and ensures the development of new scion shoots that may be needed for trunk renewal the following season.

A common practice in the northeastern United States is the use of multiple trunk vines. In this area most *V. vinifera* vines have three to five trunks per vine, and renewals are brought up each year to replace trunks that were killed by the cold and/or crown gall. Although this practice does not eliminate the pathogen from the vineyard, it helps to ensure a crop and to manage the disease at a tolerable level.

Because *A. tumefaciens* is systemic within vines, and because biovar 3 has not been detected in nonvineyard soils, the planting of pathogen-free vines may be effective in controlling crown gall and in preventing the high incidences of the disease that have often been experienced in new plantings. The significance of soil inoculum in replanted vineyards is not known.

Selected References

Burr, T. J., and Katz, B. H. 1983. Isolation of *Agrobacterium tumefaciens* biovar 3 from grapevine galls and sap, and from vineyard soil. Phytopathology 73:163–165.

Burr, T. J., and Katz, B. H. 1984. Grapevine cuttings as potential sites of survival and means of dissemination of *Agrobacterium tumefaciens.* Plant Dis. 68:976–978.

Burr, T. J., Bishop, A. L., Katz, B. H., Blanchard, L. M., and Bazzi, C. 1987a. A root-specific decay of grapevine caused by *Agrobacterium tumefaciens* and *A. radiobacter* biovar 3. Phytopathology 77:1424–1427.

Burr, T. J., Katz, B. H., and Bishop, A. L. 1987b. Populations of *Agrobacterium* in vineyard and nonvineyard soils and grape roots in vineyards and nurseries. Plant Dis. 71:617–620.

Kerr, A., and Panagopoulos, C. G. 1977. Biotypes of *Agrobacterium radiobacter* var. *tumefaciens* and their biological control. Phytopathol. Z. 90:172–179.

Lehoczky, J. 1971. Further evidences concerning the systemic spreading of *Agrobacterium tumefaciens* in the vascular system of grapevine. Vitis 10:215–221.

Moore, L. W., Anderson, A., and Kado, C. I. 1980. *Agrobacterium.* Pages 17–25 in: Laboratory Guide for Identification of Plant Pathogenic Bacteria. N. W. Schaad, ed. American Phytopathological Society, St. Paul, MN. 72 pp.

Tarbah, F. A., and Goodman, R. N. 1986. Rapid detection of *Agrobacterium tumefaciens* in grapevine propagating material and the basis for an efficient indexing system. Plant Dis. 70:566–568.

(Prepared by T. J. Burr)

Bacterial Blight

A grapevine decline first reported in Italy in 1879 and in France in 1895 was studied for many years in these countries and was usually attributed to the bacterium *Erwinia vitivora* (Baccarini) du Plessis, a synonym of the ubiquitous saprophyte *E. herbicola* (Lohnis) Dye, which commonly exists on plant surfaces and as a secondary organism in lesions caused by many plant pathogens. The nature of the decline was confused until 1969, when the true causal agent, *Xanthomonas ampelina,* was isolated and identified in Greece.

It now has been confirmed that bacterial blight or bacterial necrosis caused by *X. ampelina* occurs in Greece, France, Spain, Italy, Portugal, Turkey, and South Africa. The disease is also known as *mal nero* in Italy; *maladie d'Oléron, nécrose bactérienne,* and *carbou* in France; *tsilik marasi* in Greece; *necrosis bacteriana* in Spain; *mal negro* in Portugal; and *vlamsiekte* in South Africa. Moreover, a very similar disease syndrome has been reported from Austria, Switzerland, Yugoslavia, Bulgaria, Tunisia, the Canary Islands, and Argentina and has been attributed to *E. vitivora.* Therefore, the disease most probably occurs in these countries as well.

Bacterial blight is a chronic, systemic disease of significant economic importance. The pathogen is endemic in several grape-growing regions where it affects commercially important cultivars. No curative control measures are known. Losses arise from reduced productivity and shortened longevity of diseased vineyards. Some vineyards have been abandoned because of their progressive deterioration. Serious damage has been reported on the very susceptible cultivar Thompson Seedless (Sultanina) in Greece and on Alicante Bouschet, Ugni blanc, Granache, and Maccabeu in France.

Symptoms

The bacteria attack the vascular tissues, causing shoot blights and cankers and occasionally leaf spots. Symptoms are most conspicuous in early spring until midsummer. In early spring, budbreak on affected spurs is either delayed or does not occur. Other spurs on the same vine grow normally. Budbreak may occur on only a few spurs, and some of the developing shoots may be stunted, weak, and/or chlorotic, with dark brown streaks on one side. These shoots eventually wilt and die. During this period, affected branches and spurs appear slightly swollen because of hyperplasia of the cambial tissues, which have a soft, cheesy consistency. Longitudinal cracks occur in the bark at the swollen regions.

The first symptoms on the young, tender shoots appear two or three weeks after the buds begin to grow. Longitudinal cracks develop, starting from the lower internodes and slowly extending upward. These internally borne cracks, which are surrounded by dark brown to black necrotic tissue, deepen, extend to the pith, and develop into cankers (Plate 76). Leaves on infected shoots may exhibit sectorial or marginal necrosis of their blades (Plate 77) and one-sided cracking of their petioles (Plate 78). Cracks and cankers may also appear on the peduncle and rachis of the clusters (Plate 78). The vascular tissues of infected shoots, canes, petioles, and tendrils often exhibit a one-sided, reddish or brown discoloration of the xylem vessels that is readily seen in longitudinal and cross sections. Brown necrotic spots 1–2 mm in diameter and usually surrounded by halos occasionally develop on young, tender leaves (Plate 77).

Symptoms may vary considerably depending on the cultivar and possibly environmental conditions. In several cultivars, shoot and cane cracks and cankers do not develop or do so only occasionally. It is not unusual for chronically diseased vines that show little sign of disease to suddenly develop acute symptoms and become debilitated. Affected vines may be scattered but more often occur in pockets in the vineyard.

The symptoms of bacterial blight can be confused with those of several other diseases, including Phomopsis cane and leaf spot, Eutypa dieback, fanleaf degeneration, grapevine yellows, shoot necrosis, corky bark, boron deficiency, and *Rotbrenner*. Therefore, laboratory identification of the causal agent is essential. Standard bacteriological tests can be used, but the indirect immunofluorescence technique and sensitivity to phage ϕ 15 allow rapid identification. The bacterium may also be detected rapidly in grapevine sap using the indirect immunofluorescence technique.

Causal Organism

X. ampelina Panagopoulos is a gram-negative, aerobic rod with one polar flagellum. It grows very slowly. On Difco nutrient agar, at 26° C, it forms round, entire, glistening, pale yellow colonies 0.2–0.3 mm in diameter in six days; up to 15 days' incubation may be required before colonies appear that are 1 mm in diameter.

X. ampelina is an atypical member of the genus *Xanthomonas*. It is readily distinguished from the four other species of the genus by its failure to grow at 33° C and to produce mucoid growth. It also does not produce acid from glucose or alkali from sodium propionate or hydrolyze esculin. It is positive in urease activity, uses tartrate, and produces a brown, diffusible pigment on yeast extract-galactose-chalk agar. A new name for this bacterium, *Xylophilus ampelinus* (Panagopoulos 1969) Williams et al, has recently been suggested.

Disease Cycle and Epidemiology

X. ampelina attacks only grapevine and survives in the vascular tissues of infected plants and cuttings. Bacteria move in xylem vessels in late winter and spread transvascularly into healthy branches and spurs and later into the new, developing shoots and grape clusters. Cankers developing on new growth provide inoculum for direct infection through the stomata of leaves during wet weather or when overhead sprinkler irrigation is used in early spring. However, the vine sap that exudes from pruning wounds of diseased plants is the most important source of inoculum.

The bacteria are disseminated locally by contaminated tools during pruning and/or by rain, especially in windy weather. They may enter healthy plants through pruning or frost wounds. Transmission of the pathogen through soil and roots during the flooding treatment for control of phylloxera has been reported in France. The bacterium is transmitted over long distances and into noninfested regions by the planting or grafting of contaminated material, which may be symptomless. Up to 50% of apparently healthy canes obtained from diseased vineyards have been found to be infected.

Plant tissues are most susceptible to infection, at least in Crete (Greece), from November to late January and are least susceptible in February and March. Prolonged wet weather, overhead irrigation, and flooding contribute to disease outbreaks.

Control

Control efforts should be directed toward preventing the spread of the pathogen to unaffected grape-growing regions and in newly established vineyards. All planting and grafting material should be obtained from disease-free areas, and all nursery stock should be inspected and handled using proper sanitation procedures. In Australia all introductions of grapevine propagation material are treated with hot water and grown in quarantine for inspection, to avoid the inadvertent introduction of the disease.

In already contaminated vineyards, all infected branches and canes should be removed by pruning and burned, and all dead or severely infected plants should be uprooted and burned. In Greece it is recommended that pruning be done in dry weather and as late as possible in the dormant season. In France the first pruning should be done in mid-January and the second as late as possible. Pruning tools should be disinfected between vines. Sprays with Bordeaux mixture or fixed copper are recommended immediately after pruning and periodically up to the stage of half-expanded leaves; they appear to be especially helpful in rainy areas. Overhead sprinkler irrigation should be avoided.

Selected References

Bradbury, J. F. 1973. *Xanthomonas ampelina.* Descriptions of Pathogenic Fungi and Bacteria, No. 378. Commonwealth Mycological Institute, Kew, Surrey, England.

Bradbury, J. F. 1984. Genus II. *Xanthomonas* Dowson 1939, 187. Pages 199–210 in: Bergey's Manual of Systematic Bacteriology. Vol. 1. N. R. Krieg, and J. G. Holt, eds. The Williams & Wilkins Co., Baltimore. 964 pp.

Grasso, S., Moller, W. J., Refatti, E., Magnano di San Lio, G., and Granata, G. 1979. The bacterium *Xanthomonas ampelina* as causal agent of a grape decline in Sicily. Riv. Patol. Veg. 15:91–106.

Lopez, M. M., Gracia, M., and Sampayo, M. 1987. Current status of *Xanthomonas ampelina* in Spain and susceptibility of Spanish cultivars to bacterial necrosis. Bull. OEPP/EPPO Bull. 17:231–236.

Matthee, F. N., Heyns, A. J., and Erasmus, H. D. 1970. Present position of bacterial blight (vlamsiekte) in South Africa. Deciduous Fruit Grower 20:81–84.

Panagopoulos, C. G. 1969. The disease tsilik marasi of grapevine: its description and identification of the causal agent (*Xanthomonas ampelina* sp. nov.). Ann. Inst. Phytopathol. Benaki 9:59–81.

Ride, M., and Marcelin, H., eds. 1983. La nécrose bactérienne de la vigne (*Xanthomonas ampelina*). Bulletin Technique des Pyrenees-Orientales, No. 106. Institut National de la Recherche Agronomique, Paris. 87 pp.

Willems, A., Gillis, M., Kersters, K., Van Den Broecke, L., and De Ley, J. 1987. Transfer of *Xanthomonas ampelina* Panagopoulos 1969 to a new genus, *Xylophilus* gen. nov., as *Xylophilus ampelinus* (Panagopoulos 1969) comb. nov. Int. J. Syst. Bacteriol. 37:422–430.

(Prepared by C. G. Panagopoulos)

Pierce's Disease

Pierce's disease is a principal factor limiting production of both *V. labrusca* and *V. vinifera* grapes in the Gulf Coastal Plains of the United States. In California it kills *V. vinifera* grapevines in isolated, discrete areas, which are called "hotspots" in that state.

The disease was first described in 1892 from the Santa Anna River Valley near Anaheim in southern California. Several decades later, the disease was found in Florida and other areas of the southeastern United States. It was subsequently identified in Mexico, Costa Rica, and Venezuela, and it likely occurs in most areas of Central America and southward into the northern parts of South America, where the causal bacteria are present and potential insect vectors are abundant. It has never been confirmed to occur in countries outside the Americas.

Symptoms

Symptoms vary with the species and cultivar that is affected. Symptoms in muscadine and other native American grapes from the southeastern United States are milder than those in *V. vinifera.* Symptoms are usually more pronounced in vines that are stressed by high temperatures and droughty conditions.

Chlorotic spots develop on leaf blades near the point of initial infection. The discoloration intensifies, and the surrounding tissues begin to wither and dry. The spots gradually enlarge. Starting near the margin of the leaf blade, tissues become completely desiccated and die. In late summer drying spreads inward in concentric zones until the entire blade may be affected (Plate 79). Leaves then often drop from the vine at the point of attachment to the petiole, leaving the petiole still attached to the shoot (Plate 80). Symptoms develop in adjacent leaves along the shoot both above and below the point of initial infection. Flower clusters on infected vines may set berries, but these usually dry up.

Late in the season, wood on affected canes fails to mature normally, leaving green "islands" of tissue surrounded by dark brown, mature wood (Plate 80). These islands persist into the dormant season and can be seen on canes throughout the winter, or until the cane is pruned or dies from frost. Tips of shoots often die the first year the vine is infected. Initially, only one or a few canes on a vine show foliar and wood symptoms.

In chronically affected vines, budbreak in spring is delayed by as much as two weeks. The new shoots grow slowly and are stunted. The first four to six leaves that form on new shoots are small, and the tissues immediately along major veins appear dark green against a chlorotic background (Plate 81). Subsequent leaves are normal in color but small. Internodes of such shoots are much shorter than normal.

Suckers growing from the base of chronically affected vines frequently appear to develop normally. Such apparent recovery may persist until the middle or end of summer; then the characteristic leaf and wood symptoms reappear on most leaves, shoots, and canes.

An affected vine may die the first year after infection (Plate 82), or it may continue to live for five or more years, depending on the species and cultivar, the age of the vine when infected, and local climatic conditions. In the American tropics, infected *V. vinifera* cultivars usually die within 12 months of planting.

Causal Organism

When Pierce's disease was first scientifically investigated during the late 1930s and for a considerable time thereafter, a virus was believed to be the causal agent. Experiments conducted during the early 1970s, however, showed that antibiotic treatments suppressed the symptoms and that immersing vines in hot water eliminated the causal agent. Subsequent studies with electron microscopy demonstrated the presence of rickettsialike bacteria in xylem elements of diseased vines in both Florida and California. In 1978 a bacterium was cultured from infected vines and Koch's postulates were completed, proving that this bacterium was the cause of the disease.

The causal agent is a small, nutritionally fastidious, gram-negative bacterium with a convoluted cell wall of several layers and extracellular fibrous strands. It has recently been named *Xylella fastidiosa* Wells et al.

Disease Cycle and Epidemiology

X. fastidiosa is widely distributed in native plants throughout its natural range in the Americas. Natural hosts occur among both monocotyledonous and dicotyledonous plant families. In California, wild grasses, sedges, and lilies are frequent hosts in hotspot areas, but native forbs, bushes, and trees are also often infected. The bacterium also causes diseases in almonds and alfalfa in California and in macadamia nuts in Costa Rica. In wild host plants as well as in grapevines and other cultivated hosts, the bacterium survives and proliferates in xylem tissues.

In the Americas, many genera of sharpshooter leafhoppers (Cicadellidae) and spittlebugs (Cercopidae) can serve as vectors of the bacterium as they feed on the xylem tissues of host plants. Such insects typically suck up large quantities of xylem fluids during feeding and may ingest bacteria when feeding on either an infected grapevine or an alternative host of the bacterium.

Bacteria attach to the foregut ("mouth") of the insect. Infective vector insects egest fluid while feeding on xylem tissues of healthy plants, and bacteria are probably transmitted with this fluid. The frequency of transmission of the bacterium during the feeding process is high for some of the vector species.

Disease symptoms appear as the bacteria increase into dense aggregates in the xylem vessels. These bacterial aggregates, along with tyloses and gums produced by the grapevine, plug the plant's vascular elements, restricting water conduction to tissues. The bacteria also produce a phytotoxin that may contribute to symptom development.

In Florida the disease spreads mostly from vine to vine in the vineyard. Sharpshooter leafhoppers on wild host plants apparently are rarely infective. In California the disease spreads mostly from weedy host plants near the margin of the vineyard and thus is common among plants growing within the first 100 m from a hotspot location.

Control

Within the natural range of *X. fastidiosa* and its insect vectors, *V. vinifera* and *V. labrusca* grapes become infected and die quickly. Only muscadine and *Euvitis* grape cultivars that are developed from grapes that occur naturally within the normal range of the bacterium will survive. The use of resistant cultivars is the only effective control for Pierce's disease in the Gulf Coastal Plains states, as well as along the east coast of Mexico and in the American tropics. Quarantine procedures for excluding the disease are irrelevant within the native range of the causal bacterium. Pierce's disease is a case where sensitive crop plants may be introduced to locations within the natural range of the pathogen rather than vice versa.

In California and many areas in states north and west of the Gulf of Mexico, *X. fastidiosa* occurs only in isolated hotspots, and delimitation of these hotspots and site selection away from such areas are effective in controlling the disease. The range of the bacterium in wild vegetation extends from northern California southward in the western United States. The disease extends southward from about the latitude of Tennessee in the

1. Powdery mildew fungus on the surface of a leaf of the cultivar Carignane. (Courtesy J. Schlesselman)

2. Distortion and stunting of young Carignane leaves affected by powdery mildew. (Courtesy J. Schlesselman)

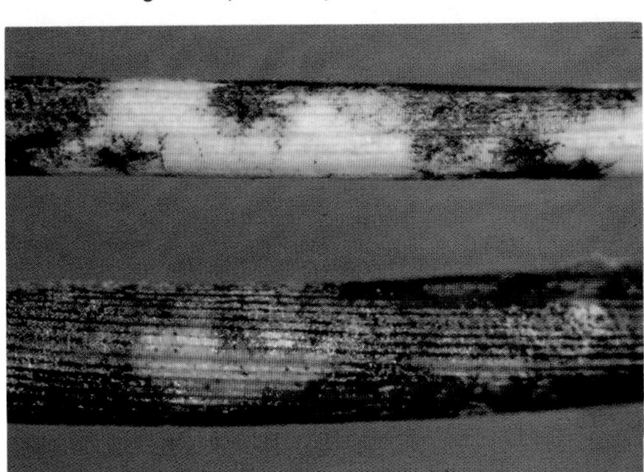

3. Blackened tissue on shoots infected with *Uncinula necator*. (Courtesy R. C. Pearson)

4. Blotchy ripening of Chancellor berries infected with *Uncinula necator*. (Courtesy R. C. Pearson)

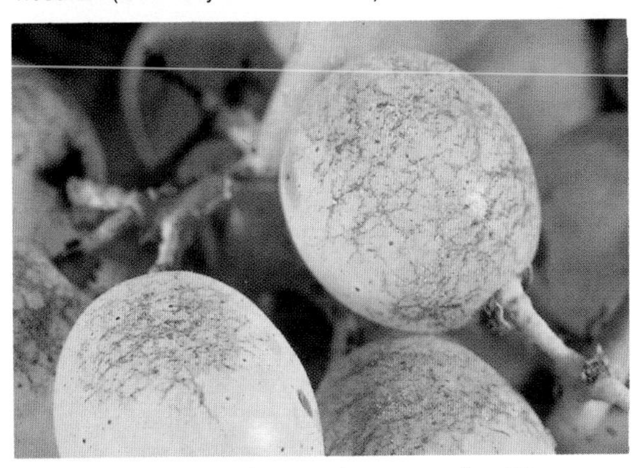

5. Netlike pattern of scar tissue on ripe Thompson Seedless (Sultanina) berries infected with *Uncinula necator*. (Courtesy J. Schlesselman)

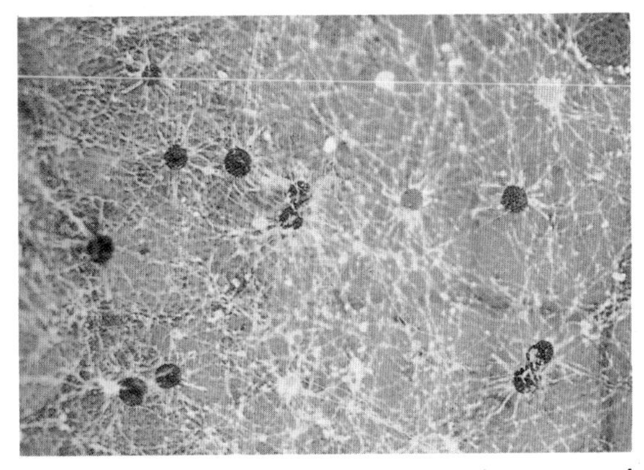

6. Cleistothecia of *Uncinula necator* in various stages of maturity. (Courtesy W. Gärtel)

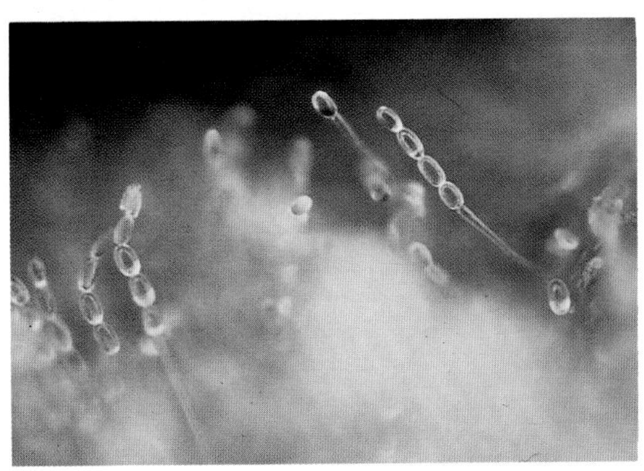

7. Conidia of *Uncinula necator*, formed in chains. (Courtesy W. Gärtel)

8. Stunted shoot covered with mycelium of *Uncinula necator* (flag shoot) that grew from an infected bud. (Courtesy R. C. Pearson)

9. "Oil-spot" symptom of downy mildew on a leaf in spring. (Courtesy R. Lafon)

10. Downy mildew symptoms on a leaf in autumn. (Courtesy R. Lafon)

11. Sporulation of *Plasmopara viticola* on the underside of a diseased leaf. (Courtesy R. Lafon)

12. "Shepherd's crook" symptom of downy mildew on a shoot in spring. (Courtesy R. C. Pearson)

13. Symptoms of downy mildew on a young cluster of grapes (gray rot). (Courtesy R. Lafon)

14. Sporulation of *Plasmopara viticola* on a young berry. (Courtesy W. Gärtel)

15. Symptoms of downy mildew on a developed cluster of grapes (brown rot). (Courtesy R. Lafon)

16. *Botrytis cinerea* on a leaf of Müller-Thurgau. (Courtesy R. C. Pearson)

17. *Botrytis cinerea* on inflorescences. (Courtesy J. Bulit)

18. Withering of grapes due to rachis infection (stalk rot) by *Botrytis cinerea*. (Courtesy J. Bulit)

19. Botrytis bunch rot. (Courtesy B. Dubos)

20. Storage rot caused by *Botrytis cinerea*. (Courtesy R. C. Pearson)

21. Black rot lesions on a Dutchess leaf. (Courtesy R. C. Pearson)

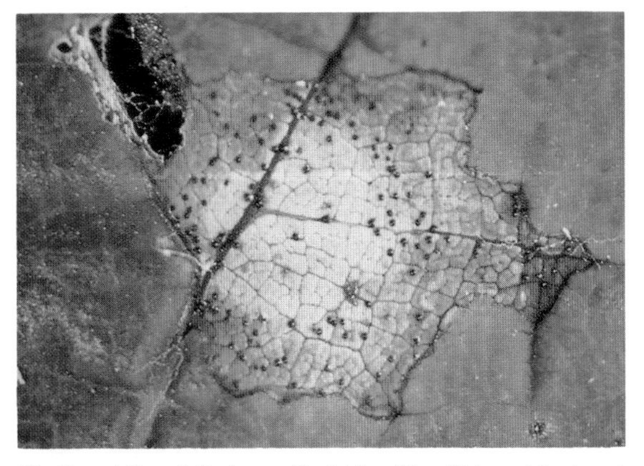

22. Pycnidia of *Guignardia bidwellii* within a black rot lesion. (Courtesy R. C. Pearson)

23. Black rot in various stages of disease development on a cluster of Aurore grapes. (Courtesy R. C. Pearson)

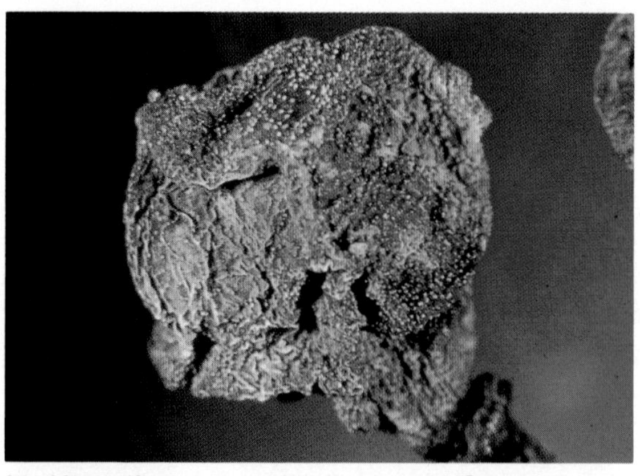

24. Mummified berry with black rot. (Courtesy L. V. Madden)

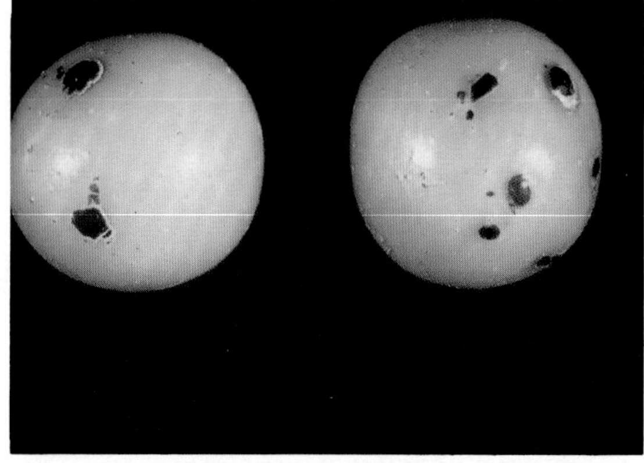

25. Scabby lesions of black rot on muscadine berries. (Courtesy P. Bertrand)

26. Yellowing and necrotic tissue on leaves infected with *Phomopsis viticola.* (Courtesy W. B. Hewitt)

27. Lesions of Phomopsis cane and leaf spot on a shoot and rachis, with pycnidia visible on the one-year-old cane. (Courtesy R. C. Pearson)

28. Black pycnidia on ripe fruit of Niagara infected with *Phomopsis viticola.* (Courtesy R. C. Pearson)

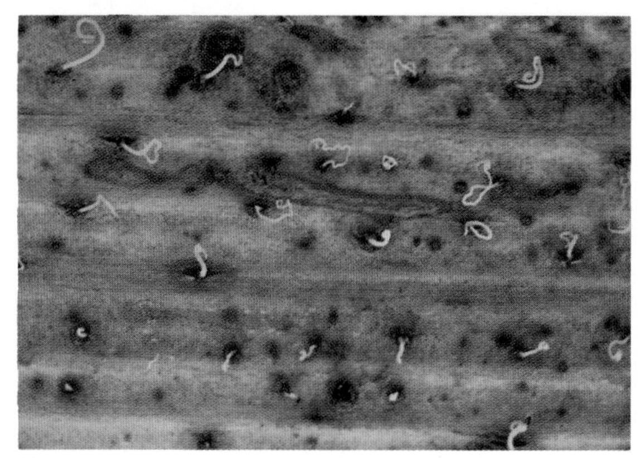

29. Pycnidia exuding spores of *Phomopsis viticola* in cirri on the surface of a one-year-old cane. (Courtesy W. Gärtel)

30. Anthracnose lesions on a leaf. (Courtesy W. Gärtel)

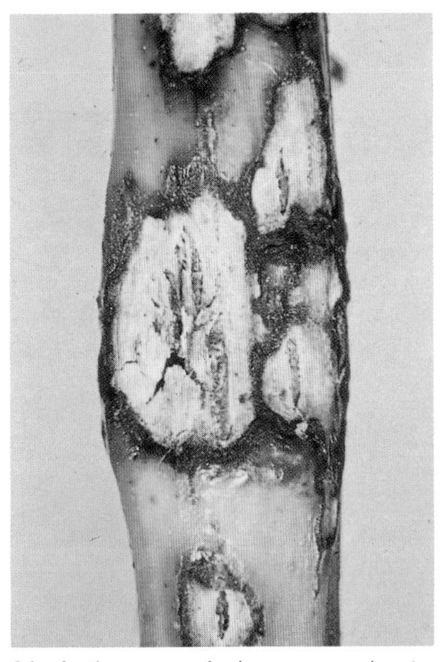

31. Anthracnose lesions on a shoot. (Courtesy W. Gärtel)

32. Anthracnose lesions on berries of Vidal blanc. (Courtesy R. C. Pearson)

33. Symptoms of *Rotbrenner* on leaves of a white cultivar. (Courtesy W. Gärtel)

34. Dried inflorescences on a cluster affected by *Rotbrenner.* (Courtesy A. Bolay)

35. Bitter rot on Aurore fruit. (Courtesy M. A. Ellis)

36. White rot on the distal and middle portions of Corvina clusters. (Courtesy Osservatorio per le Malattie delle Piante, Verona, Italy)

37. Mature pycnidia of *Coniella diplodiella* on Barbera. (Courtesy M. Bisiach)

38. Ripe rot on muscadine grapes. (Courtesy R. D. Milholland)

39. Macrophoma rot on muscadine grapes. (Courtesy P. Bertrand)

40. Angular leaf spot on muscadine grape leaves. (Courtesy P. Bertrand)

41. Diplodia cane dieback on Thompson Seedless (Sultanina) canes that were in contact with wet soil. (Courtesy W. B. Hewitt)

42. Bluish sporulation of blue mold, caused by *Penicillium* spp. (Courtesy R. C. Pearson)

43. Rhizopus rot. (Courtesy R. C. Pearson)

44. Sour bunch rot. (Courtesy J. Schlesselman)

45. Yellow uredial pustules of the grape rust fungus, *Physopella ampelopsidis*. (Courtesy L. S. Leu)

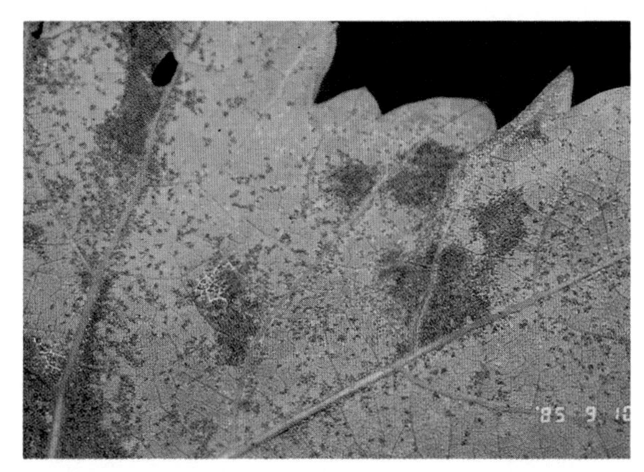

46. Brown telia of the grape rust fungus. (Courtesy L. S. Leu)

47. Pycnial lesions induced by the grape rust fungus on *Meliosma myriantha*. (Courtesy A. Kudo)

48. Aecia of the grape rust fungus on *Meliosma myriantha*. (Courtesy A. Kudo)

49. Leaf blotch (caused by *Briosia ampelophaga*) on Rougeon leaf. (Courtesy J. R. McGrew)

50. Zonate leaf spot, caused by *Cristulariella moricola*. (Courtesy J. R. McGrew)

51. Septoria leaf spot on C 3309. (Courtesy R. C. Pearson)

52. Rupestris speckle on *V. rupestris*. (Courtesy R. C. Pearson)

53. Weak, stunted shoots with shortened internodes on cordon of Chenin blanc with Eutypa dieback. (Reprinted, by permission, from Moller and Kasimatis, 1981)

54. Large canker of Eutypa dieback surrounding an old pruning wound on a trunk with the rough outer bark removed. (Reprinted, by permission, from Moller and Kasimatis, 1981)

55. Wedge-shaped zone of necrotic sapwood, exposed in cross section, indicating the extent of invasion by *Eutypa lata.* (Reprinted, by permission, from Moller and Kasimatis, 1981)

56. Cavities of perithecia of *Eutypa lata* revealed in a shallow slice cut from the stroma with a sharp blade. (Reprinted, by permission, from Moller and Kasimatis, 1981)

57. Foliar symptoms of esca on a white cultivar. (Courtesy W. Gärtel)

58. Symptoms of black measles on Thompson Seedless (Sultanina) berries. (Courtesy J. Schlesselman)

59. Sudden death (apoplexy) of part of a vine due to esca. (Courtesy B. Dubos)

60. Necrotic heartwood of a vine affected with esca. (Courtesy B. Dubos)

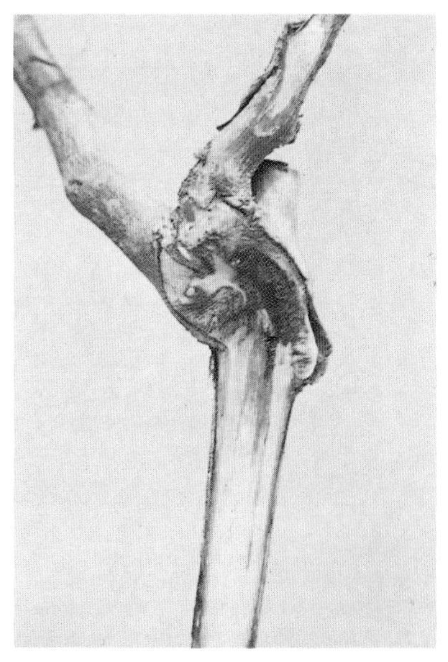

61. Black discoloration in the xylem at the graft union of Red Traminer, caused by the black dead arm fungus. (Courtesy J. Lehoczky)

62. Mycelial mat of *Armillaria mellea* beneath bark. (Courtesy R. D. Raabe)

63. Rhizomorphs of *Armillaria mellea*. (Courtesy R. D. Raabe)

64. Mushrooms of *Armillaria mellea*. (Courtesy R. D. Raabe)

65. Dead leaves adhering to a diseased grapevine, a diagnostic symptom of Phymatotrichum root rot. (Courtesy D. Téliz)

66. Mycelial strands of *Phymatotrichum omnivorum* on the surface of dead infected root, a reliable diagnostic indicator of Phymatotrichum root rot. (Courtesy Department of Plant Pathology, CAE La Laguna CIAN-INIA-SARH, Matamoros, Coah., Mexico)

67. Discolored xylem in the trunk of a Chenin blanc vine infected with *Verticillium dahliae*. (Courtesy R. C. Pearson)

68. Wilted leaves and dying shoots on part of a Semillon vine infected with *Verticillium dahliae*. (Courtesy A. C. Goheen)

69. Hyphae of *Dematophora necatrix* growing from an infected plant under moist conditions. (Courtesy A. H. McCain)

70. Small, discontinuous plaques of *Dematophora necatrix* on an infected root with bark removed. (Courtesy R. D. Raabe)

71. Canker of darkened wood on a *V. vinifera* vine, caused by *Phytophthora megasperma,* revealed by cutting into the tissue with a knife. (Photo by A. N. Kasimatis; reprinted, by permission, from Moller, 1981)

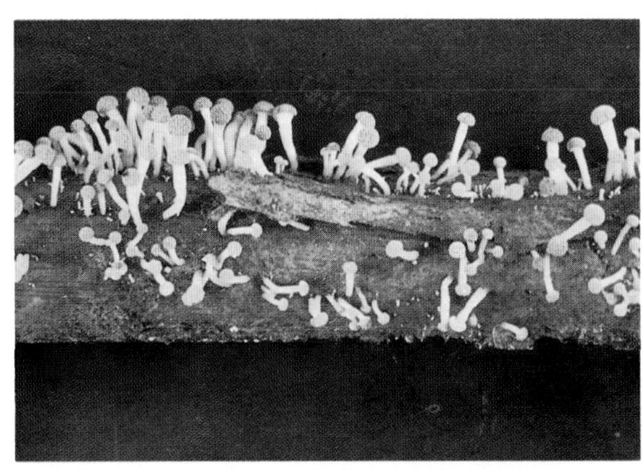

72. Apothecia of *Roesleria hypogaea* on a partially rotted root of *V. vinifera*. (Courtesy W. Gärtel)

73. Crown gall, which is most apparent on trunks near or just below the soil line and extending upward. (Courtesy T. J. Burr)

74. Small, pimplelike galls of crown gall extending up a trunk. (Courtesy T. J. Burr)

75. Crown gall at the graft union of a grapevine in a commercial nursery. (Courtesy T. J. Burr)

76. Symptoms of bacterial blight at the base of a young shoot, including a deep canker. (Courtesy W. Gärtel)

77. Leaf spots and marginal necrosis caused by *Xanthomonas ampelina*. (Courtesy C. G. Panagopoulos)

78. Necrosis and cracking of petioles and rachis due to bacterial blight. (Courtesy A. Bolay)

79. Concentric "burning" from the margin toward the point of attachment to the petiole of a leaf from a Merlot vine with Pierce's disease. (Courtesy A. C. Goheen)

80. Cane from a Merlot vine with Pierce's disease, showing islands of green tissue surrounded by periderm and petioles remaining after the leaf blades have dropped. (Courtesy A. C. Goheen)

81. Leaves on a new shoot on a Pinot noir vine with Pierce's disease, showing dark green tissues along the major veins against a chlorotic background. (Courtesy A. C. Goheen)

82. Dying Pinot noir vine with Pierce's disease. (Courtesy A. C. Goheen)

83. Shoots of Baco blanc with weeping posture and rolled, golden leaves typical of *flavescence dorée.* (Courtesy A. Caudwell)

84. Shoot of White Riesling affected by *Vergilbungskrankheit,* showing a lack of lignification and black pustules arranged in longitudinal rows. (Courtesy W. Gärtel)

85. Stunted shoot of Baco blanc with necrotic growing point, overlapping leaves, and snakelike appearance typical of *flavescence dorée.* (Courtesy A. Caudwell)

86. Angular and creamy spots along the main veins and necrosis developing in late summer on a leaf of Baco blanc affected by *flavescence dorée.* (Courtesy A. Caudwell)

87. Cluster of Baco blanc affected by *flavescence dorée.* (Courtesy A. Caudwell)

88. Healthy leaf of French Colombard (left) compared with a leaf infected with grapevine fanleaf virus (right). (Reprinted, by permission, from Raski et al, 1983)

89. Proliferation of shoots at a node on a *V. vinifera* vine infected with grapevine fanleaf virus. (Courtesy G. P. Martelli)

90. Healthy fruit of Cabernet Sauvignon (left) compared with fruit infected with grapevine fanleaf virus (right). (Reprinted, by permission, from Raski et al, 1983)

91. Yellow mosaic symptom on Thompson Seedless (Sultanina) leaves infected with grapevine fanleaf virus. (Reprinted, with permission, from Raski et al, 1983)

92. Patchy distribution of vines with yellow mosaic symptoms of fanleaf degeneration. (Courtesy G. P. Martelli)

93. Veinbanding symptoms on a Cabernet Sauvignon leaf infected with grapevine fanleaf virus. (Courtesy A. C. Goheen)

94. Oak leaf patterns on Baco noir leaves infected with tomato ringspot virus. (Courtesy D. Gonsalves)

95. Cascade berry clusters from healthy vines (left) and from vines infected with tomato ringspot virus (right). (Courtesy J. K. Uyemoto)

96. Short, crooked internodes associated with peach rosette mosaic virus infection. (Courtesy D. C. Ramsdell)

97. Berry shelling resulting from peach rosette mosaic virus infection. (Courtesy D. C. Ramsdell)

98. Chardonnay vine affected by leafroll (left), with leaves yellowing prematurely, slightly smaller than a healthy vine of the same age (right). (Courtesy A. C. Goheen)

99. Leafroll-affected Pinot noir leaf in the fall, showing downward rolling and reddening between the major veins, while the major veins remain green. (Courtesy A. C. Goheen)

100. Healthy fruit cluster (left) and leafroll-affected cluster (right) of Queen at harvest. (Courtesy A. C. Goheen)

101. Cabernet Sauvignon vine with corky bark; affected leaves do not drop from the vine until several days after the first killing frost. (Courtesy A. C. Goheen)

102. Deep grooves in the wood of a Chardonnay scion on AXR 1 rootstock with corky bark, exposed when the bark is removed from the trunk. (Courtesy A. C. Goheen)

103. Cross-sectional cut on a St. George rootstock from under a vine with corky bark. (Reprinted, by permission, from Goheen, 1981)

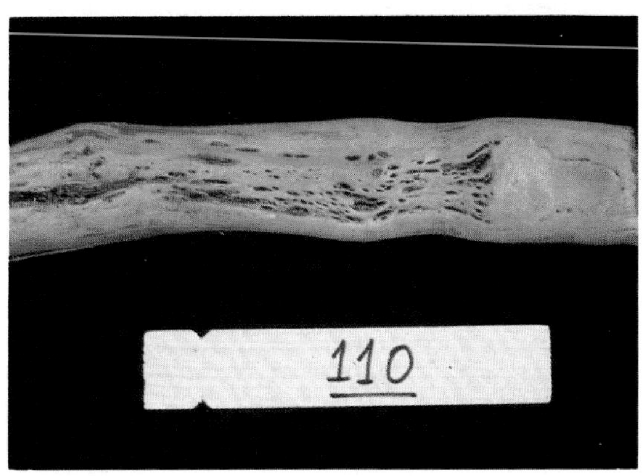

104. Symptoms of rupestris stem pitting on St. George, showing a line of small pits developing basipetally from the point of inoculation. (Courtesy A. C. Goheen)

105. Young, developing leaf on a St. George vine with fleck disease. (Courtesy G. Stellmach)

106. Irregular chlorotic spots and veinbanding on a leaf blade from a vine with vein mosaic. (Courtesy W. Gärtel)

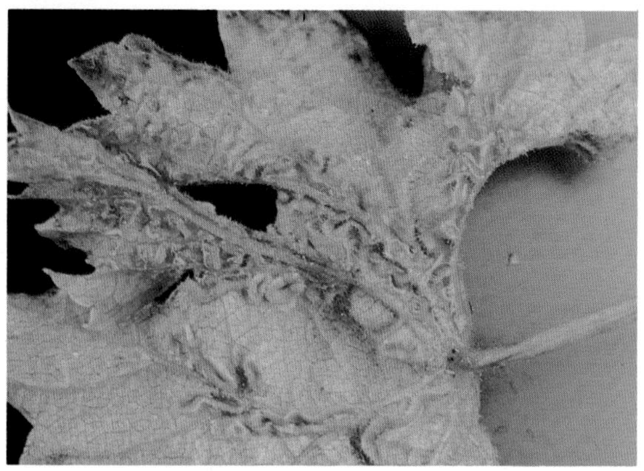

107. Outgrowths (enations) on the underside of a basal leaf from a vine with grape enation. (Courtesy W. Gärtel)

108. Chlorotic speckled bands caused by yellow speckle disease on a leaf of the cultivar Mission. (Courtesy A. C. Goheen)

109. Roots of Cabernet Sauvignon damaged by root-knot nematodes, showing galls and swellings on secondary rootlets. (Courtesy D. J. Raski)

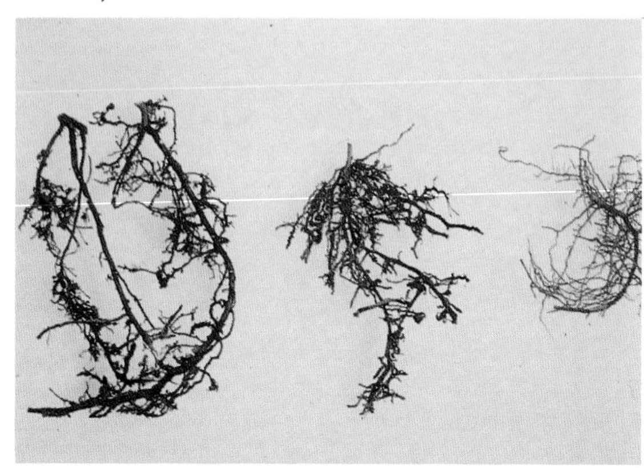

110. Roots damaged by dagger nematodes (left and center) and healthy roots (right). (Courtesy D. J. Raski)

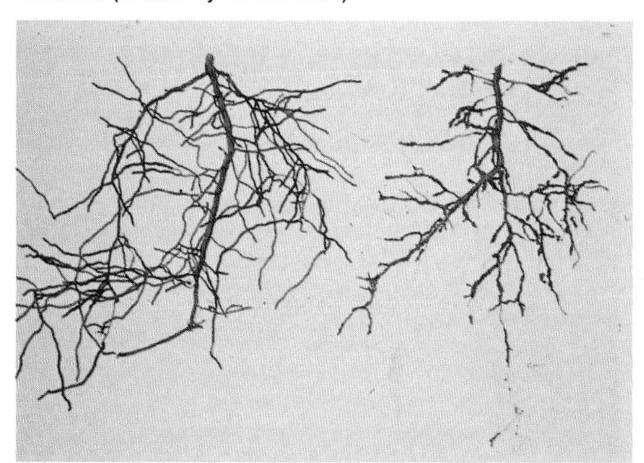

111. Roots damaged by citrus nematodes (right) and healthy roots (left). (Courtesy D. J. Raski)

112. Typical yellowing caused by Willamette spider mites in foliage of a white cultivar. (Reprinted, by permission, from Flaherty et al, 1981)

113. Leaf burn caused by Pacific spider mites on the top of a grapevine. (Reprinted, by permission, from Flaherty et al, 1981)

114. Symptoms of leafroll (left) and Willamette spider mite injury (right) on a dark cultivar. (Courtesy A. N. Kasimatis)

115. European red mite injury on foliage. (Courtesy J. A. Cox)

116. Erinea on the lower surface of a leaf, caused by the grape erineum mite. (Reprinted, by permission, from Kido, 1981)

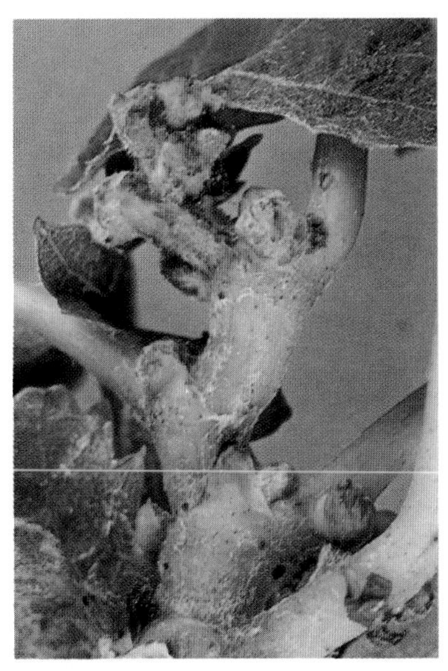

117. Stunted, zigzag growth and lateral shoot development caused by the bud mite strain of the grape erineum mite on *V. vinifera* in early spring. (Courtesy W. Gärtel)

118. Symptoms of the leaf curl strain of the grape erineum mite on French Colombard. (Courtesy F. Jensen)

119. Darkened blotches caused by false spider mites (*Brevipalpus chilensis*) on a Ribier rachis in Chile. (Courtesy R. H. Gonzalez)

120. Typical starfish pattern of berry scarring caused by western flower thrips on Thompson Seedless (Sultanina). (Photo by A. N. Kasimatis; reprinted, by permission, from Jensen et al, 1981)

121. European grape thrips injury to White Malaga berries. (Reprinted, by permission, from Jensen et al, 1981)

122. European grape thrips injury to Thompson Seedless (Sultanina) foliage in Chile. (Courtesy R. H. Gonzalez)

123. Injury caused by eastern grape leafhoppers on Concord leaves. (Courtesy G. L. Jubb, Jr.)

124. Injury caused by potato leafhoppers on Catawba leaves. (Courtesy G. L. Jubb, Jr.)

125. Angular reddening produced by *Empoasca* leafhopper feeding on Alden leaves. (Courtesy A. C. Goheen)

126. Pinot noir shoot girdled by treehopper feeding, showing red leaves distal to the feeding site. (Courtesy R. C. Pearson)

127. Nodosities caused by the root form of phylloxera on Concord roots. (Courtesy R. N. Williams)

128. Phylloxera galls on the underside of a leaf. (Courtesy R. N. Williams)

129. Variegation chimera on leaves of *V. vinifera*. (Courtesy A. C. Goheen)

130. Variegation chimera in fruit of Muscat blanc. (Courtesy A. C. Goheen)

131. Genetic fasciation on Petite Sirah (Durif) vine. (Courtesy A. C. Goheen)

132. Sauvignon blanc shoot with leaf distortion chimera (bottom) and normal shoot (top). (Courtesy A. C. Goheen)

133. Chlorotic leaf margins and sporadic necrotic spots, spring symptoms of potassium deficiency. (Courtesy W. Gärtel)

134. Marginal and interveinal necrosis accompanied by downward rolling of leaf margins, advanced spring symptoms of potassium deficiency. (Courtesy W. Gärtel)

135. Black leaf symptom of potassium deficiency, seen in late summer but only on leaves that receive direct sunlight. (Courtesy W. Gärtel)

136. Magnesium-deficient leaf, showing the green margins and interveinal chlorosis that distinguish magnesium deficiency from deficiencies of potassium or other nutrients. (Courtesy W. Gärtel)

137. Severe magnesium deficiency symptoms; older leaves show the most advanced expression. (Courtesy W. Gärtel)

138. Stem necrosis (*Stiellähme*). (Courtesy W. Gärtel)

139. Symptoms of *Säureschäden*, which is found on vines growing in soils of very low pH. (Courtesy W. Gärtel)

140. Interveinal chlorosis and necrosis caused by iron deficiency, which is usually associated with soil high in lime or, less often, with cold, wet soil. (Courtesy R. M. Pool)

141. Manganese deficiency symptoms, showing interveinal chlorosis beginning as chlorotic islands. (Courtesy W. Gärtel)

142. Leaf with interveinal chlorosis, reduced size, blade asymmetry, and opened petiolar sinus caused by zinc deficiency. (Courtesy W. Gärtel)

143. Loose clusters of seeded berries that vary in size, from a zinc-deficient vine. (Courtesy W. Gärtel)

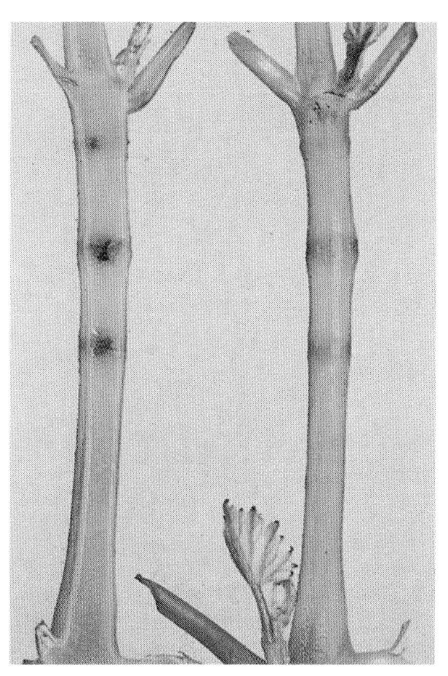

144. Longitudinal section of an internode (left), showing internal necrosis due to boron deficiency, and external symptoms (right). (Courtesy W. Gärtel)

145. Necrotic shoot tip, interveinal chlorosis of leaves (in advanced stages, interveinal necrosis), and swellings on internodes, caused by boron deficiency. (Courtesy W. Gärtel)

146. Cluster with one normal seeded berry and many small, seedless berries of equal size, caused by boron deficiency. (Courtesy W. Gärtel)

147. Chenin blanc leaves with marginal necrosis caused by excess boron. (Courtesy L. P. Christensen)

148. Cabernet Sauvignon vine with drought stress symptoms. (Courtesy A. N. Kasimatis)

149. Heat-injured (left) and uninjured (right) shoot tips of Chenin blanc. (Courtesy L. P. Christensen)

150. Heat injury to leaves and clusters of Pinot noir. (Courtesy A. N. Kasimatis)

151. Heat injury (sometimes called Almeria spot) to Thompson Seedless (Sultanina) berries. (Courtesy A. N. Kasimatis)

152. Dried and separated pith of a grape shoot struck by lightning. (Courtesy W. Gärtel)

153. Petite Sirah vine with symptoms of heat stress and damage to vine shoots after a lightning strike, which splintered the stake. (Courtesy A. N. Kasimatis)

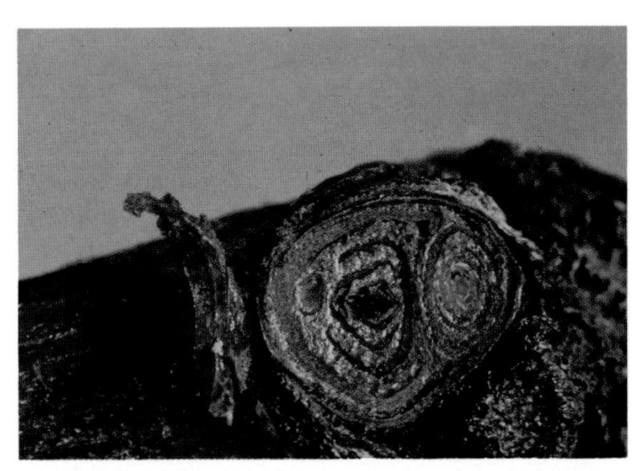

154. Cross section of a winter-injured node of Dutchess, showing the dead primary shoot bud. (Courtesy T. J. Zabadal)

155. Irregular growth from White Riesling buds following the death of some primary buds from winter cold. (Courtesy R. M. Pool)

156. Malformed and mottled leaf developing after cold injury. (Courtesy R. C. Pearson)

157. Discoloration of phloem tissue due to low-temperature injury at the base of a two-year-old Chardonnay trunk, with healthy xylem. (Courtesy R. C. Pearson)

158. Midsummer collapse of a Pinot noir vine following winterkill of the trunk base, with prolific growth of ground suckers. (Courtesy R. M. Pool)

159. Spring freeze injury to growing Thompson Seedless (Sultanina) shoots. (Courtesy A. N. Kasimatis)

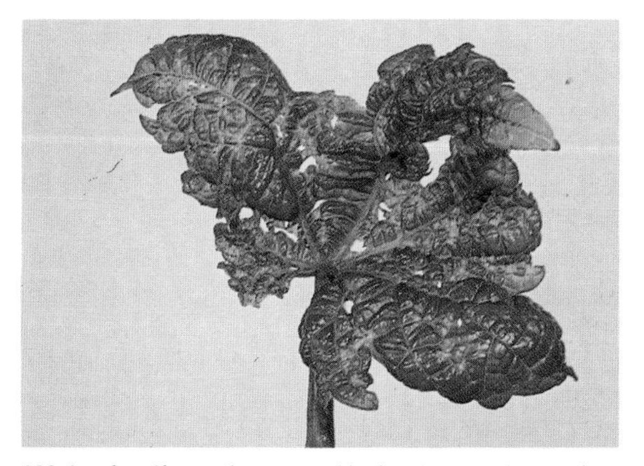

160. Leaf malformation caused by low temperatures when buds were swollen but before budbreak. (Courtesy W. Gärtel)

161. Hail injury to a young Thompson Seedless (Sultanina) shoot. (Courtesy A. N. Kasimatis)

162. Injury from blowing sand, on young shoots and leaves of Thompson Seedless (Sultanina). (Courtesy A. N. Kasimatis)

163. Salt toxicity on Chardonnay, with marginal necrosis developing into general leaf scorch. (Courtesy A. N. Kasimatis)

164. Oxidant stipple on the upper surface of leaves of *V. vinifera*. (Courtesy L. H. Weinstein)

165. Marginal and interveinal necrosis induced by fluoride on leaves of *V. vinifera* in Germany. (Courtesy W. Gärtel)

166. Marginal and interveinal chlorosis and necrosis induced by sulfur dioxide on leaves of *V. labrusca*. (Courtesy R. C. Musselman)

167. Shoot injured by glyphosate. (Courtesy A. N. Kasimatis)

168. Simazine injury. (Courtesy R. M. Pool)

169. Diuron injury to Concord leaves. (Courtesy R. C. Pearson)

170. 2,4-D injury to Flame Tokay shoots. (Courtesy A. N. Kasimatis)

171. Dicamba injury to Thompson Seedless (Sultanina) foliage. (Courtesy A. N. Kasimatis)

172. Aminotriazole injury to a Thompson Seedless (Sultanina) leaf. (Courtesy A. N. Kasimatis)

173. Paraquat injury to a Petite Sirah leaf. (Courtesy A. N. Kasimatis)

174. Leaf injured by translocated paraquat. (Courtesy A. N. Kasimatis)

175. Effect of excess gibberellic acid on White Riesling grapes. (Courtesy R. C. Pearson)

176. Shoots of Chenin blanc one year after treatment with gibberellic acid, showing very small clusters (leaves were removed to expose the fruit). (Courtesy A. N. Kasimatis)

177. Sulfur injury to foliage of Sauvignon blanc. (Courtesy A. N. Kasimatis)

178. Sulfur injury to berries. (Courtesy A. N. Kasimatis)

179. Sulfur dioxide damage to Emperor grapes in storage, characterized by bleached and sunken tissue. (Courtesy J. Harvey)

180. Severe copper sulfate injury to Concord foliage. (Courtesy R. C. Pearson)

181. Stunted, malformed leaves on Aurore shoot injured by dinocap. (Courtesy R. C. Pearson)

182. Dinocap injury associated with exposure to the sun, on a Concord leaf. (Courtesy R. C. Pearson)

183. Etaconazole injury to Aurore foliage, showing stunted lateral shoots with small, downward-cupped leaves. (Reprinted, by permission, from Pearson, 1986; copyright 1986 American Chemical Society)

184. Marginal and interveinal chlorosis on Elvira leaves injured by benalaxyl. (Courtesy R. C. Pearson)

185. Leaf injured by vinclozolin. (Courtesy W. Gärtel)

186. Berries injured by captan. (Courtesy W. Gärtel)

187. Node injured by sodium arsenite. (Courtesy L. P. Christensen)

188. Chancellor leaves injured by endosulfan. (Courtesy G. L. Jubb, Jr.)

eastern states. Pierce's disease is not a problem where the bacterium is not established in the wild.

Broad-spectrum tetracycline antibiotics have been somewhat effective in protecting grapevines against Pierce's disease in the southeastern United States, but achieving adequate control by chemicals on a commercial scale does not seem feasible. In California, antibiotic treatments of vines growing in hotspots have not been successful.

Insect trapping on sticky boards, the capture and identification of potential vector insects in sweep nets, and serological detection of the bacteria in wild host plants and in macerated insects have helped to delimit hotspots. Insecticidal treatments to control vectors in hotspot areas have not been effective in eliminating the disease.

Quarantine measures for preventing disease spread to areas outside the natural range of the bacterium are probably unnecessary. Cuttings or propagation buds from affected vines do not survive long enough to establish the disease in a new area. To ensure the health of plants taken to new areas, however, propagating wood can be immersed in water at 45° C for 3 hr. Dormant grape cuttings readily survive this treatment, which kills any Pierce's disease bacteria that are present in the wood.

Selected References

Alderz, W. C., and Hopkins, D. L. 1979. Natural infectivity of two sharpshooter vectors of Pierce's disease of grape in Florida. J. Econ. Entomol. 72:916–919.

Davis, M. J., Purcell, A. H., and Thomson, S. V. 1978. Pierce's disease of grapevines: Isolation of the causal bacterium. Science 199:75–77.

Goheen, A. C., Nyland, G., and Lowe, S. K. 1973. Association of a rickettsialike organism with Pierce's disease of grapevines and alfalfa dwarf and heat therapy of the disease in grapevines. Phytopathology 63:341–345.

Hopkins, D. L. 1983. Gram-negative, xylem-limited bacteria in plant disease. Phytopathology 73:347–350.

Hopkins, D. L., and Mollenhauer, H. H. 1973. Rickettsia-like bacterium associated with Pierce's disease of grapes. Science 179:298–300.

Wells, J. M., Raju, B. C., Hung, H. Y., Weisburg, W. G., Mandelco-Paul, L., and Brenner, D. J. 1987. *Xylella fastidiosa* gen. nov., sp. nov.: Gram-negative, xylem-limited, fastidious plant bacteria related to *Xanthomonas* spp. Int. J. Syst. Bacteriol. 37:136–143.

(Prepared by A. C. Goheen and D. L. Hopkins)

Grapevine Yellows Diseases

Flavescence Dorée

Flavescence dorée, originally known as *maladie du Baco 22 A,* was first described in 1956 in Gascony in southwestern France. From there it spread rapidly east to southern France, northern Italy, and possibly Slovenia and Rumania.

Symptoms

In spring, the growth of newly infected susceptible vines is usually inhibited: budbreak is delayed or absent, internodes are progressively shortened, and certain parts of the leaves atrophy. The characteristic symptoms appear in summer, when in the most sensitive cultivars, the vine often adopts a weeping posture, the shoots bending down as though made of rubber (Plate 83). Lignification is absent throughout the whole shoot, and black pustules sometimes are arranged in longitudinal rows along the shoot (as in *Vergilbungskrankheit*) (Plate 84). Growing points rapidly become necrotic (Plate 85).

The leaves harden, roll slightly abaxially, and tend to overlap, giving the shoot a characteristic snakelike appearance (Plate 85). The brittle leaves first become golden yellow in white cultivars (Plate 83) and red in black cultivars on all parts (lamina and veins) most exposed to the sun. Later in summer, "creamy" spots appear along the main veins (Plate 86). These cream-colored spots generally become necrotic. Sometimes angular spots occur, whose spread is limited by two or three main veins. These angular spots are yellow in white cultivars and red in black-fruited cultivars.

If the symptoms appear before or during bloom, the entire inflorescence dries up. If the symptoms appear later in the season, the peduncle dries and darkens (Plate 87), and the berries shrivel and develop a dense, fibrous, and bitter pulp that makes the fruit unusable.

The symptoms of *flavescence dorée* and of other grapevine yellows diseases are sometimes confused with those of certain virus diseases, chiefly leafroll and corky bark and the yellow mosaic syndrome of fanleaf degeneration. The most characteristic symptoms of all of the grapevine yellows diseases are the shape and the spread of the creamy spots along the veins of the leaves and the angular spots associated with hardening and light rolling of the leaves. Leaves affected by leafroll or corky bark may roll but never develop the creamy or angular spots; leaves affected by fanleaf degeneration may have these spots but are not rolled or brittle. Another distinctive symptom of grapevine yellows diseases is the lack of lignification of the wood, often associated with black pustules. No lack of lignification occurs in leafroll, and no pustules occur in either leafroll or corky bark. The third characteristic symptom of grapevine yellows, unknown in leafroll, corky bark, and fanleaf degeneration, is the shriveling of the berries.

The phloem and the pith of affected shoots are very developed, whereas the xylem is less developed. This is particularly striking in the stem, where the annual ring of wood growth may be reduced to a one-cell layer. Phloem fibers are rare and irregular, which explains the rubberlike habits of the shoots as well as their tendency to break. The outer phloem becomes necrotic; cell walls appear yellowish, swollen, and compressed on each other, which may explain the blocking of assimilates in the leaves and the subsequent lack of maturity of fruit and wood.

Causal Organism

Flavescence dorée was first thought to be a virus disease of the "yellows type" because it could be transmitted by grafting and by leafhoppers. However, no virus or any other pathogen, including mycoplasmalike organisms (MLOs), has been found in situ in grapevine tissues by transmission electron microscopy. Nevertheless, an MLO is the suspected causal agent.

The pathogen can be transmitted from grapevine to *Vicia faba* and to *Chrysanthemum carinatum* by the leafhopper *Scaphoideus littoralis* Ball (syn. *S. titanus* Ball) and from these plants back to grapevine. MLOs are readily detected in diseased *Vicia faba* tissue and in infected leafhoppers but not in healthy plants or healthy leafhoppers.

Another leafhopper, *Euscelidius variegatus* (Kirschbaum), can be used to transmit the pathogen among *Vicia faba* plants. Furthermore, it is possible to trap MLOs from infected *E. variegatus* using an antiserum prepared from infected *Vicia faba.* Reciprocally, MLOs from infected *Vicia faba* can be trapped with an antiserum prepared from infected *E. variegatus.* This system has been used to visualize the MLO by immunosorbent electron microscopy and to detect the MLO by enzyme-linked immunosorbent assay (ELISA).

Disease Cycle and Epidemiology

In the field, the *flavescence dorée* pathogen is transmitted by the leafhopper *S. littoralis.* This leafhopper is probably native to the eastern United States and Canada and was introduced into Europe after World War II. *S. littoralis* has only one generation per year. Hatching usually begins in the second half of May in southern France, about one week after bloom on Baco blanc (Baco 22 A). The length of the hatching period (five weeks in southwestern France, 12 weeks in Corsica) depends on winter chilling, which is necessary for terminating egg diapause. There are five nymphal stages. The adults first appear in July in France and begin to lay eggs one week later. The eggs are inserted into the phloem of the woody parts of vines and into the buds. The adults disappear in early September.

The pathogen overwinters in infected canes (the same as those used for winter grafting) in an incubation phase. It is ingested in spring by *S. littoralis* (either nymph or adult), and after a three- to four-week incubation period, the insect becomes infectious.

In susceptible cultivars and in temperate climates, symptoms often appear in the year following natural inoculation by the vector. The symptoms are severe and systemic throughout the plant; this has been termed "the crisis year." However, in some vines infected with other viruses, such as tomato black ring virus, the symptoms of *flavescence dorée* are not systemic but remain localized around the inoculation site.

Two types of disease reaction can develop during subsequent years, the Nieluccio type and the Baco 22 A type. In the Nieluccio type, symptoms become progressively more severe each year until the vine dies. Most cultivars (Baco blanc, Ugni blanc, Grenache, Baroque, Colombard, Jurancon, Aramon, and others) exhibit the Baco 22 A type reaction, which is recovery. If not reinoculated, they grow the following spring, without expressing symptoms. Recovery is complete, and the wood loses all its infectivity. If such cultivars are reinoculated during the "crisis year" or within the next two to three years, only localized symptoms appear on a few shoots around the inoculation point. Several such inoculations, each producing localized symptoms, may mimic systemic infection. After four to five years, this defense reaction of the plant disappears; thereafter, reinoculation again leads to systemic symptoms. Only cultivars capable of recovery develop localized symptoms. Localized symptoms and recovery may be caused by the same plant defense reaction.

Flavescence dorée can be transmitted experimentally by dormant grafting. For this purpose, diseased canes are unusable, and the disease cannot be transmitted from recovered ones. Only canes inoculated during the preceding summer, and thus in a state of incubation during winter, can be used. This is a major problem because such canes do not show symptoms and can be mistakenly identified as healthy. Wood for graft transmission studies should be collected from vines in the vicinity of infected plants in vineyards where natural infection is high.

Control

Flavescence dorée epidemics are associated with the presence of the leafhopper *S. littoralis*. In Europe, the distribution of the insect greatly exceeds the regions in which the disease occurs, but it does not include all the regions where the insect is expected to be able to proliferate. Thus, some regions face two threats: the importation of the insect and of the pathogen. After both have been introduced, the disease spreads very quickly, either around foci in the vineyard or by means of flying adult leafhoppers helped by wind to new areas up to 30 km away. The number of infected vines can increase sevenfold within one year.

It is important to prevent importation of the vector's eggs, which are occasionally present in the bark of commercial rootstocks. In regions where the insect is already present, caution is necessary in handling canes that may carry the disease in the incubation phase. Imported vines should be kept in quarantine nurseries, protected from flying insects, for one or two years. Alternatively, warm water treatment of cuttings (72 hr in agitated water at 30° C) eliminates the pathogen.

Once the disease is introduced in an area, it becomes endemic and constitutes a permanent risk. It is then advisable to cultivate commercially desirable resistant cultivars if they are available. Until this solution is implemented, it is necessary to protect sensitive vines from *S. littoralis*.

To control the disease by insecticidal treatments, insecticides should be applied during the egg hatching period. In practice, it is sufficient to apply the first treatment three weeks after the first observed hatch, because the insect is not infectious during its incubation period. Depending on the length of the hatching period and insecticide retention, three (southwestern France) or six (Corsica) treatments are necessary. An alternative method of insect control is to kill the eggs before budbreak by dilute sprays of oleoparathion.

Some species of *Vitis*, such as the American species *V. labrusca* and *V. rupestris*, are not susceptible to *flavescence dorée*. Some hybrids, such as Couderc 13, are susceptible only when very young.

Selected References

Caudwell, A. 1964. Identification d'une Nouvelle Maladie à Virus de la Vigne, la Flavescence Dorée. Etude des Phénomènes de Localisation des Symptômes et de Rétablissement. Ann. Epiphyt. 15 (Hors Série I). Institut National de la Recherche Agronomique, Paris. 197 pp.

Caudwell, A. 1983. L'origine des jaunisses à mycoplasmes (MLO) des plantes et l'exemple des jaunisses de la vigne. Agronomie 3:103–111.

Caudwell, A., and Schvester, D. 1970. Flavescence dorée. Pages 201–207 in: Virus Diseases of Small Fruits and Grapevines (a Handbook). N. W. Frazier, ed. Division of Agricultural Sciences, University of California, Berkeley. 290 pp.

Caudwell, A., Kuszala, C., Bachelier, J. C., and Larrue, J. 1970. Transmission de la flavescence dorée de la vigne aux plantes herbacées par l'allongement du temps d'utilisation de la cicadelle *Scaphoideus littoralis* Ball et l'étude de sa survie sur un grand nombre d'espèces végétales. Ann. Phytopathol. 2:415–428.

Caudwell, A., Moutous, G., Brun, P., Larrue, J., Fos, A., Blancon, G., and Schick, J. P. 1974. Les épidémies de flavescence dorée en Armagnac et en Corse et les nouvelles perspectives de lutte contre le vecteur par des traitements ovicides d'hiver. Minist. Agric. Bull. Tech. Inf. 294:783–794.

Caudwell, A., Meignoz, R., Kuszala, C., Schneider, C., Larrue, J., Fleury, A., and Boudon, E. 1982. Purification immunologique et observation ultramicroscopique en milieu liquide de l'agent pathogène (MLO) d'une jaunisse végétale, la flavescence dorée de la vigne. C. R. Seances Acad. Agric. Fr. 68:407–415.

Schvester, D., Carle, P., and Moutous, G. 1961. Sur la transmission de la flavescence dorée des vignes par une cicadelle. C. R. Seances Acad. Agric. Fr. 47:1021–1024.

(Prepared by A. Caudwell)

Bois Noir and *Vergilbungskrankheit*

Bois noir (black wood) is a grapevine yellows disease described in northeastern France (Burgundy, Jura, and Champagne) and in Switzerland. *Vergilbungskrankheit* was described in the Moselle and Rhine valleys in the Federal Republic of Germany. These diseases occur in neighboring regions of Europe, and they may actually be the same disease. Reports describing grapevine yellows in northern Italy seem to indicate that both *bois noir* and *flavescence dorée* may occur in this region. The causal agents of *bois noir* and *Vergilbungskrankheit* are not known.

The symptoms of *bois noir* and *Vergilbungskrankheit* are identical to those of *flavescence dorée*, but the diseases differ in several respects. First, the cultivars susceptible to *flavescence dorée* are different from those susceptible to *bois noir* or *Vergilbungskrankheit;* for example, Pinot noir is susceptible to *flavescence dorée* but not to *bois noir*. Second, *Scaphoideus littoralis*, the leafhopper vector of *flavescence dorée*, does not transmit *bois noir* or *Vergilbungskrankheit*. And third, the epidemiology of the diseases differs. *Flavescence dorée* generally appears with consistent severity and spreads from southwestern France to other regions. *Bois noir* and *Vergilbungskrankheit* can be either severe or mild, depending on the individual infected vine, and do not appear to spread.

Selected References

Caudwell, A. 1961. Etude sur la maladie du bois noir de la vigne: Ses rapports avec la flavescence dorée. Ann. Epiphyt. 12:241–262.

Gärtel, W. 1965. Untersuchungen über das Auftreten und das Verhalten der Flavescence dorée in den Weinbaugebieten an Mosel und Rhein. Weinberg Keller 12:347–376.

Mendgen, K. 1971. Untersuchungen über eine Vergilbungskrankheit der Reben an Rhein, Mosel und Saar. Weinberg Keller 18:345–431.

(Prepared by A. Caudwell)

Other Grapevine Yellows

An important yellows disease has been described in Sicily, chiefly on the cultivar Insolia, in the absence of *Scaphoideus littoralis.* Similar symptoms are described in Greece and in Galilee in Israel, in regions where *S. littoralis* may not be able to complete the winter diapause of its eggs because of the lack of sufficiently low temperatures. Thus, a Mediterranean grapevine yellows may exist that is not related to *flavescence dorée,* or that may be *flavescence dorée* but is transmitted in hot regions by a leafhopper other than *S. littoralis.*

Several yellows diseases have been described on grapevine in other regions, including the "Rhine Riesling problem" in Australia, now called Australian grapevine yellows, and *amarillamiento de Elqui,* in Elqui Valley in northern Chile and probably also in Argentina.

An interesting problem is the occurrence of a grapevine yellows disease in New York State. This is the original region of *S. littoralis,* where the genus *Scaphoideus* supposedly evolved. The disease has been described on *Vitis* interspecific hybrids and *V. vinifera* grown in a region where *V. labrusca* originated and where *V. labrusca* vines are traditionally cultivated.

All of the grapevine yellows diseases so far described show the symptoms of *flavescence dorée.*

Selected References

Granata, G. 1982. Deperimenti e giallume in piante di vite. Inf. Fitopatol. 32(7–8):18–20.

Magarey, P. A. 1986. Grape-vine yellows—Aetiology, epidemiology and diagnosis. S. Afr. J. Enol. Vitic. 7:90–100.

Magarey, P. A., and Wachtel, M. F. 1982. The Rhine Riesling problem—Recent findings. Aust. Grapegrower Winemaker 220:78–80.

Pearson, R. C., Pool, R. M., Gonsalves, D., and Goffinet, M. C. 1985. Occurrence of flavescence dorée-like symptoms on 'White Riesling' grapevines in New York, USA. Phytopathol. Mediterr. 24:82–87.

Rumbos, I., and Biris, D. 1979. Studies on the etiology of a yellows disease of grapevines in Greece. Z. Pflanzenkr. Pflanzenschutz 86:226–273.

Uyemoto, J. K. 1974. A new disease affecting the grapevine variety De Chaunac. (Abstr.) Proc. Am. Phytopathol. Soc. 1:146.

(Prepared by A. Caudwell)

Diseases Caused by Viruses and Viruslike Agents

Virus in a perennial woody plant such as grapes originally meant a transmissible disease for which no obvious pathogen could be discovered. The disease and the virus had a single meaning. The disease was transmissible, and its causal agent was invisible. The disease spread to new plants or to healthy plants in the population by propagation (rooted cuttings, layers, buds, grafts, etc.), by invertebrate vectors (generally insects), or by sap inoculation. No pathogen could be recovered by isolation techniques, and no organism could be found in infected tissues with the best light microscope.

Transmission by budding, grafting, and rooting cuttings was then and still is an important characteristic of viruses found in perennial plants. Insects and especially nematodes proved to be vectors for spreading viruses, but sap was generally not important for spreading viruses in perennial plants.

With the development of electron microscopy, minute particles could be found in diseased plants. Isolation and biochemical characterization of these minute particles changed the meaning of virus. In the new meaning, the particle was the virus.

In plant pathology, Koch's postulates must be fulfilled to establish that an organism is the pathogenic agent causing a disease. These tests are difficult to perform with viruses in perennial plants such as grapes. If a purified virus particle is proved to be the causal agent, the disease is a virus disease. If no particle can be isolated and established as the causal agent, the disease remains a viruslike disease. Establishing that viruses are the cause of disease in woody plants is difficult because viruses are obligate pathogens and their pathogenicity is destroyed during purification. The cause of leafroll, an extremely important viruslike disease in grapes, has still not been definitely established.

Viruslike diseases are described in the early grape literature. Affected vines stood out in vineyards in comparison to healthy ones. No pathogenic agents causing the diseases could be found, although the diseases were perpetuated by asexual propagation.

The virus and viruslike diseases in grapes spread slowly in nature. Many would be of no economic importance if they did not spread by propagation.

Historically, viruslike diseases in grapes were named for symptoms noted on a single grape species or cultivar. Neither the causal virus nor the natural vector was known. Symptoms on one cultivar were infrequently correlated with those on a different cultivar; as a result, two or more names were often applied to the same disease.

The virus and viruslike diseases of grapevines have one feature in common: they are transmissible by grafting an affected bud to a healthy plant during plant propagation. Because of this property, diseases may be tested and compared on a single host. Sensitive healthy plants can be selected from grape populations and used as test plants. Inoculated test plants can serve as indicators of the viruslike disease. The disease can then be studied and characterized without knowing the specific pathogen or the way it spreads in nature.

Such tests for disease characterization do not require the isolation of the pathogen, but they do require sensitive indicator plants. In California and elsewhere, careful selection for freedom from disease (sanitary selection) was used to select test plants that showed the presence or absence of specific disease symptoms when they were inoculated by buds from a plant of unknown health status. Healthy plant lines can be established by reciprocally grafting buds from different selections until one is found that does not transmit symptoms to others when grafted but does show symptoms when an affected plant is grafted to it. The test plants for inoculating are called indicators; the test itself is called indexing.

Fanleaf degeneration, tomato ringspot virus decline, tobacco ringspot virus decline, and peach rosette mosaic virus decline are important grape virus diseases caused by known viruses. Leafroll, corky bark, and rupestris stem pitting are important

TABLE 2. Symptoms[a] on Indicator Plants that Separate Vines with Leafroll, Corky Bark, and Rupestris Stem Pitting from Healthy Vines

Disease Indicated	Indicator Plant		
	Cabernet franc	LN-33	St. George
None (healthy)	Green leaves; smooth wood	Green leaves; smooth wood	Green leaves; smooth wood
Leafroll	Red leaves with green veins; smooth wood	Either red leaves with green veins or green leaves; smooth wood	Green leaves; smooth wood
Corky bark	Red leaves with green veins; smooth wood	Uniformly red or bronze leaves; grooved or grooved and pitted wood	Green leaves; grooved wood
Rupestris stem pitting	Green leaves; smooth wood	Green leaves; smooth wood	Green leaves; pitted wood along stem below inoculation point

[a] Leaf color and pitting or grooving that develops on the woody cylinder of chip bud-inoculated indicator plants 18 months after inoculation.

viruslike diseases that are recognized on the basis of comparisons made in standard indicators. Each of these diseases damages plants and causes crop loss. Recognition of the disease and, even more important, recognition of healthy mother plants are essential for controlling both virus and viruslike diseases in grape.

Fanleaf degeneration is caused by grapevine fanleaf virus, which is transmitted in the soil by nematodes. Tomato ringspot, tobacco ringspot, and peach rosette mosaic viruses are also transmitted by nematodes in the soil, and each causes a decline disease in grapevines. These nematode-transmitted viruses have the same particle morphology; they are polyhedral particles approximately 30 nm in diameter. For this reason they are all in a group called nepoviruses (*ne*matode-transmitted, *po*lyhedral viruses).

The symptoms of leafroll, corky bark, and rupestris stem pitting overlap, but the diseases can be separated by grafting an unknown disease source to the indicators Cabernet franc, St. George, and LN-33. The leaf and woody cane reactions in the indicators separate the three diseases (Table 2). The indicators can also be used to establish clean mother lines of grape cultivars, because no reaction in any indicator means freedom from disease.

Recently, three separate viroids were found in grapevines. Their significance is unknown. Their presence has not yet been correlated with any of the virus or viruslike diseases.

Selected References

Bovey, R., Gärtel, W., Hewitt, W. B., Martelli, G. P., and Vuittenez, A. 1980. Virus and Viruslike Diseases of Grapevines. Payot, Lausanne; La Maison Rustique, Paris; and Verlag Eugen Ulmer, Stuttgart. 181 pp.

Frazier, N. W., ed. 1970. Virus Diseases of Small Fruits and Grapevines (a Handbook). Division of Agricultural Sciences, University of California, Berkeley. 290 pp.

Semancik, J. S., Rivera-Bustamante, R., and Goheen, A. C. 1987. Widespread occurrence of viroid-like RNAs in grapevines. Am. J. Enol. Vitic. 38:35–40.

(Prepared by A. C. Goheen)

Fanleaf Degeneration

Fanleaf degeneration is the oldest known virus disease of *V. vinifera.* Its name comes from the peculiar malformation of infected leaves, which exhibit widely open petiolar sinuses and abnormally gathered primary veins, which give the leaf the appearance of an open fan (Plate 88). In the European literature, records of the disease date back some 200 years, and grapevine leaves with typical symptoms were found in herbaria established before the introduction of American rootstock hybrids. The consensus is that fanleaf degeneration may have existed in the Mediterranean Basin and the Near East since the earliest time of grape cultivation. Now the disease is known to occur worldwide. The disease is often called fanleaf, but the more correct name is fanleaf degeneration.

The impact of fanleaf degeneration varies with the tolerance of the cultivar to the virus. Tolerant cultivars produce fairly good crops. Sensitive cultivars are severely affected, showing progressive decline of the vines, low yields (up to 80% losses) and low fruit quality, shortened productive life of the vineyard, low proportion of graft take, reduced rooting ability of propagation material, and decreased resistance to adverse climatic factors.

Symptoms

The disease is characterized by three distinct syndromes evoked by different reactions to the causal agent.

In the first syndrome, *infectious malformations,* leaves are variously and severely distorted, asymmetrical, and puckered and show acute denticulations (Plates 88 and 89). Chlorotic mottling may sometimes accompany foliar deformations. Shoots are also malformed, showing abnormal branching (Plate 89), double nodes, short internodes, fasciations, and zigzag growth. Bunches are smaller and fewer than normal, ripen irregularly, and have shot berries and poor fruit set (Plate 90). Foliar symptoms develop early in the spring and persist through the vegetative season, although they may be somewhat masked in summer.

In the second syndrome, *yellow mosaic,* affected vines develop bright chrome yellow discolorations early in the spring that may affect all vegetative parts of the vine (leaves, shoots, tendrils, and inflorescences). Chromatic alterations of the leaves vary from a few scattered yellow spots, sometimes appearing as rings or lines, to extended mottling of the veins and/or interveinal areas, to total yellowing (Plate 91). In spring, affected plants in a vineyard are easily spotted from a distance (Plate 92). The foliage and shoots show very little if any malformation, but clusters are small, with some shot berries. In hot climates, summer vegetation resumes its normal green color, and the yellowed old growth turns whitish and tends to fade away.

In the third syndrome, *veinbanding,* chrome yellow flecking is first localized along the main veins of mature leaves and then spreads a little way into the interveinal areas (Plate 93). These discolorations appear in mid to late summer, usually in a limited number of leaves. Discolored leaves show little malformation. Fruit set is poor, clusters are straggly, and the yield may be virtually zero.

Trabeculae, or endocellular cordons, are a highly characteristic internal symptom of vines affected by fanleaf degeneration. These are radial bars crossing the lumen of epidermal, parenchyma, phloem, and xylem cells. They are made up of a pectic core surrounded by a cellulose sheath encrusted with lignin, suberin, or cutin, depending on the tissue in which they form. These structures are readily visible in lignified shoots, especially in the basal internodes, and are useful for diagnostic purposes.

Causal Agent

Grapevine fanleaf virus (GFLV) is a member of the nepovirus group. Particles are isometric and about 30 nm in diameter, with an angular outline and poorly resolved surface structure. The genome is bipartite, being composed of single-stranded RNA occurring as two functional species (with molecular weights of 1.4 and 2.4 $\times 10^6$), which are both needed for infectivity. The coat protein is a single polypeptide with a molecular weight of 54,000, encoded in the smaller RNA species.

GFLV is readily transmitted by sap inoculation but not through grapevine seeds. The virus is seed-transmitted in some experimental herbaceous hosts and occurs in their pollen and in the pollen of grapevine.

The natural host range of the virus is strictly limited to *Vitis* spp. The experimental host range is moderately wide, comprising over 30 species in seven botanical families. *Chenopodium amaranticolor* Coste and Reyn., *C. quinoa* Willd., *Gomphrena globosa* L., and *Cucumis sativus* L. are useful diagnostic species, as well as *V. rupestris* 'St. George,' when inoculated by grafting.

GFLV isolates are antigenically uniform, including those from widely separated geographical areas. A single natural serological variant has recently been found. All GFLV isolates are serologically rather distantly related to Arabis mosaic virus (ArMV). Serological diagnosis can be done by conventional gel double-diffusion test, enzyme-linked immunosorbent assay (ELISA), or immunoelectron microscopy.

Ultrastructural modifications of GFLV-infected cells of both experimental and natural hosts conform to those typically induced by nepoviruses, that is, vacuolate-vesiculate cytoplasmic inclusions, cell wall outgrowths, virus-containing tubules, and paracrystalline aggregates of virus particles.

Disease Cycle and Epidemiology

Virus-Vector Relationship. GFLV is transmitted from grape to grape by the longidorid nematodes *Xiphinema index* Thorne & Allen and *X. italiae* Meyl. A single brief feeding on an infected plant is sufficient to make nematodes viruliferous. *X. index* acquires GFLV from roots of infected vines and retains it for up to eight months in the absence of host plants or up to three months when the nematode feeds on virus-immune host plants. The virus is retained in the vector in the cuticular lining of the esophageal lumen. Adults and juvenile stages of both *X. index* and *X. italiae* transmit the virus experimentally, but in nature *X. index* seems to be by far the more efficient vector.

Virus Dissemination and Survival. GFLV cannot be disseminated over long distances by natural means because of the limited range of vector movement. Long-distance spread is achieved chiefly by transfer of infected propagation material. Its disastrous dissemination throughout the world started in the late 1800s, when phylloxera-resistant American rootstocks were introduced into Europe. It is believed that in Europe, GFLV had previously been confined with its vector(s) within scattered enclaves in the traditional grape-growing areas.

Short-range spread depends on nematodes, which, because of their limited mobility, are not efficient agents of virus dispersal. For instance, the yellow mosaic strain of GFLV spreads in the field no more than 1.3–1.5 m/year. Furthermore, although it is pollenborne, GFLV is not transmitted through grape seeds and has no natural weed hosts. Thus the only natural reservoir for this virus is the grapevine itself. Because the roots of grapevines remain viable for many years after removal of the mother plants, any rootlet that contains GFLV and can support *X. index* constitutes a source of inoculum during the interval between the uprooting of one vineyard and the establishment of the next.

Control

Several lines of action can be pursued for controlling fanleaf degeneration.

Control of Nematode Vectors. Nematode vectors cannot be successfully controlled in established vineyards. Steps must therefore be taken in the preplanting stage to break the ecological cycle of the nematode-virus complex through prolonged fallow and weed control or to eradicate vector populations with soil fumigants. Fumigation at high rates (e.g., 1,000 L of fumigant per hectare) considerably slows the rate of virus recontamination in shallow soils but not in deep soils.

Cross Protection. Experiments in France using GFLV to protect grapes against superinfection by ArMV and vice versa were encouraging, for protected vines acquired a high degree of immunization against the challenging virus. However, cross protection has many limitations as a practical control measure and is not applicable to serologically unrelated viruses.

Breeding for Resistance. Resistance to GFLV and to *X. index* has been identified, respectively, in some eastern *V. vinifera* cultivars and in *Muscadinia* or *Vitis* species other than *V. vinifera* and is being used to produce a new generation of complex rootstocks. Two such rootstocks have recently been released from the breeding program at the University of California, Davis. Scions grafted to these new rootstocks have remained disease-free even when planted in soil infested with viruliferous nematodes.

Selection and Production of Virus-Free Stocks. Sanitary selection combined with heat therapy is a powerful tool for reducing the incidence of fanleaf degeneration in newly established vineyards. Healthy stocks, when planted in nematode-free soil or in soils with populations of nonviruliferous vectors, remain healthy for the productive life of the vineyard. Vineyards planted to GFLV-free stocks are very homogeneous in morphology and productivity, the yields are improved by 40–70%, the berries contain more sugar, and wine made from the berries is satisfactory. GFLV-free material is readily obtained through conventional or slightly modified heat therapy, micrografting, or in vitro meristem and shoot-tip culture.

Selected References

Bovey, R. 1982. Control of virus and virus-like diseases of grapevine: Sanitary selection and certification, heat therapy, soil fumigation and performance of virus-tested material. Pages 299–309 in: Proc. 7th Meeting, International Council for the Study of Viruses and Virus Diseases of the Grapevine (ICVG), Niagara Falls, Canada, 1980. A. J. McGinnis, ed. Vineland Research Station, Vineland Station, Ontario. 355 pp.

Bovey, R., Brugger, J. J., and Gugerli, P. 1982. Detection of fanleaf virus in grapevine tissue extracts by enzyme-linked immunosorbent assay (ELISA) and immune electron microscopy (IEM). Pages 259–275 in: Proc. 7th Meeting, International Council for the Study of Viruses and Virus Diseases of the Grapevine (ICVG), Niagara Falls, Canada, 1980. A. J. McGinnis, ed. Vineland Research Station, Vineland Station, Ontario. 355 pp.

Hewitt, W. B., Goheen, A. C., Raski, D. J., and Gooding, G. V. 1962. Studies on virus diseases of grapevine in California. Vitis 3:57–83.

Hewitt, W. B., Martelli, G. P., Dias, H. F., and Taylor, R. H. 1970. Grapevine fanleaf virus. Descriptions of Plant Viruses, No. 28. Commonwealth Mycological Institute and Association of Applied Biologists, Kew, Surrey, England.

Martelli, G. P. 1978. Nematode-borne viruses of grapevine, their epidemiology and control. Nematol. Mediterr. 6:1–27.

Quacquarelli, A., Gallitelli, D., Savino, V., and Martelli, G. P. 1976. Properties of grapevine fanleaf virus. J. Gen. Virol. 32:349–360.

Raski, D. J., Goheen, A. C., Lider, L. A., and Meredith, C. P. 1983. Strategies against grapevine fanleaf virus and its nematode vector. Plant Dis. 67:335–339.

Vuittenez, A. 1970. Fanleaf of grapevine. Pages 217–228 in: Virus Diseases of Small Fruits and Grapevines. N. W. Frazier, ed. Division of Agricultural Sciences, University of California, Berkeley. 290 pp.

(Prepared by G. P. Martelli and V. Savino)

Tomato Ringspot Virus Decline

Tomato ringspot virus decline of grapevines occurs in the eastern United States and the Lake Ontario region of Canada and to a much lesser extent in California. The disease has not been reported in grapevines outside North America. The disease is also called virus-induced grapevine decline, grape yellow vein disease, tomato ringspot disease, and little berry disease.

Symptoms

Symptoms differ significantly in different geographical areas. In general, the disease symptoms are more severe in colder regions.

In the northeastern United States and Canada, during the first year of infection, plants produce normal growth except for a few shoots that may have leaves that are mottled and have distinct oak leaf patterns (Plate 94). These leaf symptoms are not always present throughout the growing season, which makes it very difficult to visually detect the disease in the first year. The disease is much more evident in the second year. New growth is generally weak and sparse because many infected buds are killed or injured by low winter temperatures. Infected plants are much more susceptible to winter injury than healthy ones. Shoots are short, with distinctly shortened internodes, distorted leaves, and a leaf area only one-third the normal size. The leaves may develop oak leaf patterns, but this symptom is not diagnostic because it occurs sporadically and is seen only for a short part of the growing season. The diagnostic symptoms are shortened internodes, small and distorted leaves, and stunted plants. Berries on fruit clusters are sparse and develop unevenly (Plate 95), thus resulting in a significant reduction in yield. In extremely susceptible cultivars such as Cascade, fruit production is uneconomical in the second year. In the third year, new growth is very stunted and is limited to suckers that arise near the base of the trunk, where shoot buds are less susceptible to winter injury. These suckers have shortened internodes and small leaves, and the fruit clusters are as described above. Infected vines often survive more than three years but remain stunted and unproductive (see Tobacco Ringspot Virus Decline).

In California and Maryland, the growth of vines is much less affected by the disease than in the colder Northeast and Canada. In Maryland, the virus is associated with the "little grape" disease of Vidal blanc. Infected Vidal blanc vines show no obvious decline in vine growth and no foliage symptoms, but individual berries of fruit clusters are about one-third the size of those of healthy plants. In California, vines with "grape yellow vein" disease decline slowly and foliage symptoms are evident. Leaves develop yellow flecks along the veins and also on other areas of the lamina. These symptoms vary in intensity throughout the growing season. The most consistent symptom of this disease is reduced fruit set. Individual clusters contain no berries or various numbers of normal and seedless berries.

The differences in symptom severity in these regions are probably correlated with the differences in winter temperatures, but differences in cultivars and virus strains may also contribute.

Causal Agent

Tomato ringspot virus (TomRSV) belongs to the nepovirus group and is vectored by the dagger nematodes *Xiphinema americanum, X. californicum,* and *X. rivesi.* The virus is endemic in the northeastern United States, infects a wide range of deciduous fruit crops, and is generally considered the most economically important virus infecting fruit crops in the northeastern United States.

Disease Cycle and Epidemiology

Several properties of TomRSV contribute to its rather complex biology. The virus infects a wide range of deciduous fruits, including peaches, blueberries, apples, elderberries, raspberries, cherries, and strawberries. In addition, many weeds commonly found in vineyards (e.g., dandelions, sheep sorrel, common chickweed, red clover, and narrow-leaved plantain) serve as hosts. TomRSV is seedborne in many of the weed hosts and also in grapes. The nematode vectors retain the virus for long periods of time and can acquire it from roots of infected grape or weeds.

The virus is introduced into vineyards by the planting of infected grapevines and by the dispersal of seeds from infected weeds, with subsequent spread to vines by nematodes. Because TomRSV affects the growth of grapevines significantly, especially in the Northeast, the introduction of the virus through propagation material is minimal. In Maryland, however, this method of introduction may be significant. On the other hand, introduction of the virus through seeds from infected weeds is significant in New York State. Because TomRSV survives in numerous domestic and wild fruit crops and in perennial weed plants and their seeds, it is not uncommon for the virus to be present in land that is used to establish new vineyards.

Control

The control of tomato ringspot virus decline begins with the use of virus-free cuttings for establishing vineyards. Growers should use material from certified sources whenever possible.

The disease can be controlled using resistant cultivars or rootstocks. All American (*V. labrusca*) type cultivars are resistant. The reaction of interspecific hybrids to the virus is not predictable with certainty, but Cascade, DeChaunac, Chelois, Baco noir, Vidal blanc, Ventura, Dutchess, Siegfriedrebe, and Vincent are known to be susceptible. Cultivars of *V. vinifera* are generally susceptible to TomRSV. However, only tolerant *V. vinifera* cultivars are grown in California, and the disease has virtually disappeared there.

A number of rootstocks are resistant to graft inoculation or show field resistance. Kober 5BB, St. George, 44-53 Malegue, 110 Richter, 41 B Millardet et de Grasset, Shakoka, Clinton, C 1616, and Sanona have been reported to be resistant to graft inoculation. SO 4 and C 3309 are susceptible to graft inoculation but show good field resistance, although TomRSV has been recovered from a few of these field-grown rootstocks in New York State. As expected, *V. labrusca*-type rootstocks are resistant to the virus. Resistant rootstocks control the disease when susceptible cultivars are planted in land with a history of the virus, especially in the Northeast.

Several cultural practices should be followed before planting a new vineyard and in the maintenance of established vineyards in the Northeast. Because TomRSV infects a wide range of crops and weed hosts, care should be taken to keep vineyards free of weeds, which could serve as virus reservoirs. When a new area is planted, the ground should be thoroughly worked over one year in advance of planting and weeds should be kept from reinfesting the site.

Selected References

Allen, W. R., and Van Schagen, J. G. 1982. Tomato ringspot virus in European hybrid grapevines in Ontario: A re-evaluation of the incidence and geographic distribution. Can. J. Plant Pathol. 4:272–274.

Allen, W. R., Dias, H. F., and Van Schagen, J. G. 1982. Susceptibility of grape cultivars and rootstocks to an Ontario isolate of tomato ringspot virus. Can. J. Plant Pathol. 4:275–277.

Dias, H. F. 1977. Incidence and geographic distribution of tomato ringspot virus in DeChaunac vineyards in the Niagara Peninsula. Plant Dis. Rep. 61:24–28.

Gilmer, R. M., and Uyemoto, J. K. 1972. Tomato ringspot virus in 'Baco noir' grapevines in New York. Plant Dis. Rep. 56:133–135.

Gonsalves, D. 1982. Reaction of grape varieties to tomato ringspot virus. Dev. Ind. Microbiol. 23:91–97.

Gooding, G. V., Jr., and Teliz, D. 1970. Grapevine yellow vein. Pages 238–241 in: Virus Diseases of Small Fruits and Grapevines. N. W. Frazier, ed. Division of Agricultural Sciences, University of California, Berkeley. 290 pp.

Uyemoto, J. K. 1975. A severe outbreak of virus-induced grapevine decline in Cascade grapevines in New York. Plant Dis. Rep. 59:98–101.

Uyemoto, J. K., and Gilmer, R. M. 1972. Spread of tomato ringspot virus in 'Baco noir' grapevines in New York. Plant Dis. Rep. 56:1062–1064.

Uyemoto, J. K., Martelli, G. P., Woodham, R. C., Goheen, A. C., and Dias, H. F. 1978. Grapevine (*Vitis*) virus and virus-like diseases. Set No. 1 in: Plant Virus Slide Series. O. W. Barnett and S. A. Tolin, eds. Communications Center, Clemson University, Clemson, SC. 29 pp.

(Prepared by D. Gonsalves)

Tobacco Ringspot Virus Decline

Tobacco ringspot virus (TRSV) induces decline in grapevines with symptoms that are indistinguishable from those caused by tomato ringspot virus (TomRSV). The disease has been reported only in the northeastern United States, primarily New York State and Pennsylvania. Like TomRSV, TRSV is endemic to northeastern North America. Although TRSV is serologically distinct from TomRSV and is classified as a separate virus, it belongs to the same virus group (nepovirus) and is transmitted by the same nematode vector, *Xiphinema americanum*. It is likely that *X. rivesi* is also able to transmit TRSV, although experimental evidence for this transmission has not been published. The disease cycle, epidemiology, and control measures for tobacco ringspot virus decline are identical to those described for tomato ringspot virus decline.

There are, however, several important differences between TRSV and TomRSV. Unlike TomRSV, TRSV is not a serious problem in deciduous fruit. The weed host ranges of the two viruses are very similar, with one major exception: TRSV infects plantain (*Plantago major*) but not narrow-leaved plantain (*P. lanceolata*), whereas the reverse is true for TomRSV. With few exceptions, TRSV infections have not been observed in interspecific hybrids. Instead, the disease has been confined largely to *V. vinifera* cultivars, which may account for its minor importance in the northeastern United States, where acreage of *V. vinifera* is very limited. TRSV appears to infect *V. vinifera* cultivars more readily than does TomRSV. *V. labrusca* grapevines are resistant to both viruses.

Selected References

Gilmer, R. M., Uyemoto, J. K., and Kelts, L. J. 1970. A new grapevine disease induced by tobacco ringspot virus. Phytopathology 60:619–627.

Uyemoto, J. K., Martelli, G. P., Woodham, R. C., Goheen, A. C., and Dias, H. F. 1978. Grapevine (*Vitis*) virus and virus-like diseases. Set No. 1 in: Plant Virus Slide Series. O. W. Barnett and S. A. Tolin, eds. Communications Center, Clemson University, Clemson, SC. 29 pp.

(Prepared by D. Gonsalves)

Peach Rosette Mosaic Virus Decline

Peach rosette mosaic virus (PRMV), a member of the nepovirus group, induces a decline of *V. labrusca* grapevines that occurs only in Michigan in the United States. The virus also causes a disease of peach (hence the virus name) in Michigan and in southwestern Ontario, Canada. The American cultivars Concord and Catawba are very susceptible, and Niagara and Delaware are quite resistant under field conditions. Several French-American hybrids have become infected in field tests. The disease is also called grapevine degeneration, grapevine decline, berry shelling disease, and delayed budding disease.

Vine vigor is reduced over a period of several years. Winter injury weakens diseased vines, which may die after several years. As vines become unproductive as a result of straggly clusters and berry shelling, growers remove them.

Symptoms

The usual symptom in older vineyards is a circular pattern of missing or dead vines. This pattern is typical of nepoviruses in general. Infected vines usually exhibit an umbrellalike growth habit due to shortened and crooked shoot internodes (Plate 96). Leaf deformities are common. The basal sinus is often flattened, and leaves are puckered and misshapen. Clusters are straggly, and berries shell off (Plate 97). Bud burst is sometimes, but not always, delayed.

Other disorders mimic peach rosette mosaic virus decline. Boron deficiency can cause similar shortened and crooked internodes and leaves with flattened basal sinuses. Grapevine fanleaf virus can also cause similar symptoms in *V. vinifera* cultivars and French-American hybrids.

Disease Cycle and Epidemiology

In addition to peach and grape, some perennial weeds, including dandelion (*Taraxacum officinale* Weber), Carolina horse nettle (*Solanum carolinense* L.), and curly dock (*Rumex crispus* L.), are hosts of PRMV. The nematodes *Xiphinema americanum* Cobb and *Longidorus diadecturus* Eveleigh & Allen are vectors. The virus can be endemic in weeds before a vineyard is planted, and susceptible vines may become infected by vector nematodes. Alternatively, the disease can be introduced via infected planting stock. Grape pomace containing diseased fruit that has been crushed for wine or juice may also be a source of virus introduction when spread in the vineyard to increase soil organic matter. PRMV has been shown to be in 9.5% of grape seedlings grown from seed taken from diseased vines. A low percentage (0.5%) of infectious seedborne PRMV survives after hot pressing of fruit, a process whereby the crushed grapes are held at 60° C for 2 hr. The disease spreads about 1 m/year to new vines in a circular pattern. There is a three- to four-year latent period between infection and the appearance of symptoms.

Control

Long-term fallowing of soil is not an effective means of control of peach rosette mosaic virus decline, because nematodes can remain alive and viruliferous for many years on infected surviving roots. In established vineyards, all infected vines in and around infection foci should be identified visually and by enzyme-linked immunosorbent assay (ELISA). Several vines beyond the area of infection should also be tested to ensure that all symptomless infected vines are identified. All infected vines should be removed, including crowns and major roots. The soil should be worked for several months during the summer, and the area should be fumigated in the autumn with a high rate of D-D or Vorlex (281–374 L/ha), using a superimposed shallow and deep application to the soil.

Growers who contemplate planting a new site in an area where PRMV may be present should thoroughly sample the site for vector nematodes. If they are present, a thorough fumigation is called for. Growers should plant only virus-indexed clean stock and should refrain from spreading pomace in their vineyards from processors who crush PRMV-infected grapes.

When the interspecific hybrids Aurore, Baco noir, and Vidal blanc were planted in infested soil, they became infected within 10 years. The rootstocks C 1613, C 3306, C 3309, Kober 5BB, and Riparia Gloire also became infected. In the same test, C 1202, C 1616, Teleki 5A, Chancellor, and Delaware did not become infected.

Selected References

Allen, W. R., Van Schagen, J. G., and Eveleigh, E. S. 1982. Transmission of peach rosette mosaic virus to peach, grape and cucumber by *Longidorus diadecturus* obtained from diseased orchards in Ontario. Can. J. Plant Pathol. 4:16–18.

Dias, H. F. 1975. Peach rosette mosaic virus. Descriptions of Plant Viruses, No. 150. Commonwealth Mycological Institute and Association of Applied Biologists, Kew, Surrey, England.

Dias, H. F., and Cation, D. 1976. The characterization of a virus responsible for peach rosette mosaic and grape decline in Michigan. Can. J. Bot. 54:1228–1239.

Ramsdell, D. C., and Gillett, J. M. 1985. Relative susceptibility of American, French hybrid and European grape cultivars to infection by peach rosette mosaic virus. Phytopathol. Mediterr. 24:41–43.

Ramsdell, D. C., and Myers, R. L. 1974. Peach rosette mosaic virus, symptomatology and nematodes associated with grapevine 'degeneration' in Michigan. Phytopathology 64:1174–1178.

Ramsdell, D. C., and Myers, R. L. 1978. Epidemiology of peach rosette mosaic virus in a Concord grape vineyard. Phytopathology 68:447–450.

Ramsdell, D. C., Andrews, R. W., Gillett, J. M., and Morris, C. E.

1979. A comparison between enzyme-linked immunosorbent assay (ELISA) and *Chenopodium quinoa* for detection of peach rosette mosaic virus in 'Concord' grapevines. Plant Dis. Rep. 63:74–78.
Ramsdell, D. C., Bird, G. W., Gillett, J. M., and Rose, L. M. 1983. Superimposed shallow and deep soil fumigation to control *Xiphinema americanum* and peach rosette mosaic virus reinfection in a Concord vineyard. Plant Dis. 67:625–627.

(Prepared by D. C. Ramsdell)

Leafroll

Leafroll is found in all countries where grapes are grown. Its wide distribution has come about through propagation from diseased mother vines. Symptoms do not show on all diseased vines and are not apparent during the winter season, when cuttings are made for plant propagation and distribution. Many American rootstocks show no symptoms when infected.

The disease causes chronic damage. A yield loss of 20% occurs each year for as long as diseased vines are held in a vineyard. Since leafroll does not kill vines, producers may be reluctant to remove affected vines because of the perennial nature of the grape and the cost of reestablishing vineyards.

Leafroll probably originated in the Near East along with *V. vinifera* and was carried west with grape cuttings. It does not occur naturally in wild grapes in North America. The disease that was called *rougeau* in France in the 1850s was very likely leafroll. Leafroll was called white Emperor disease and red leaf disease in early California vineyards.

Symptoms

Affected plants are slightly smaller than healthy ones (Plate 98). Leaves, shoots, canes, trunks, and root systems are slightly smaller than normal. In the spring, leaves on diseased and healthy vines appear similar, but as the season progresses, the diseased leaves turn yellowish or reddish, depending on the specific cultivar. By late summer, starting with leaves at the base of the shoot, the leaves roll downward. At this time the interveinal area of the leaf blade may be bright yellow or red, depending on the anthocyanin pigments present in the cultivar. The principal veins in the leaf remain green (Plate 99).

The disease delays fruit ripening. At harvesttime, the fruit on diseased vines is low in sugar and, especially in red or black cultivars, is pale (Plate 100). Clusters are smaller than normal, but the shape of the cluster and the size of individual berries are little changed.

Causal Agent

Particles resembling closteroviruses are frequently associated with diseased vines but have not been conclusively shown to be the cause of leafroll.

Disease Cycle and Epidemiology

No vector for the causal agent of leafroll has been established. Natural spread is slow in commercial vineyards. Frequently, diseased and healthy vines grow side-by-side in the vineyard for 40 years. Leafroll spreads when cuttings are taken from diseased scion or rootstock mother vines for propagation. Random selection of propagating wood would increase the incidence of the disease from low to high after a few propagative generations.

Control

Leafroll can be eliminated from nursery stock by indexing candidate mother vines on sensitive indicator plants. Cabernet franc is a sensitive indicator for leafroll (Table 2). If no symptoms develop in inoculated indicators after 18 months, the candidate is free of disease and can be registered as a mother vine source. Propagation from registered mother vines controls the spread of leafroll.

Selected References

Goheen, A. C. 1970. Grape leafroll. Pages 209–219 in: Virus Diseases of Small Fruits and Grapevines (a Handbook). N. W. Frazier, ed. Division of Agricultural Sciences, University of California, Berkeley. 290 pp.
Goheen, A. C., Harmon, F. N., and Weinberger, J. H. 1958. Leafroll (white Emperor disease) of grapes in California. Phytopathology 48:51–54.
Scheu, G. 1936. Mein Winzerbuch. Reichsnährstand Verlagsges. m.b.h., Berlin. 274 pp.
Stellmach, G. 1972. Die infektiöse Rollkrankheit im Hinblick auf Selektion und Erhaltungszüchtung von Rebenklonen. Dtsch. Weinbau 27:598.

(Prepared by A. C. Goheen)

Corky Bark

Corky bark is found everywhere grapes are grown, although it is less frequent than leafroll. It has been confused with *legno riccio*, a disease that occurs in Italy. The two diseases have many of the same symptoms, but when they are compared closely, *legno riccio* seems to be more complex, exhibiting both grooves and small pits in the wood. *Legno riccio* may be a dual infection of corky bark and rupestris stem pitting.

Symptoms

Symptoms of corky bark on leaves resemble those of leafroll but are usually more severe. In autumn, affected leaves become rolled and turn uniformly red or yellow, including tissues along the major veins. Leaves may not drop but may remain attached to the cane for several days after frost occurs (Plate 101). The wood of some cultivars shows deep grooves when bark is removed from the trunk (Plate 102).

Spindle shoot, once considered a separate disease, is a symptom pattern that develops in early spring on French Colombard vines with corky bark disease. As buds on this cultivar break in spring, shoots elongate rapidly but leaves on the shoots remain small, giving the vine a spindly appearance. Leaves that develop on the same shoots a few weeks later are normal in size, and as growth continues, the spindle shoot symptom disappears. A few other cultivars show the same symptom.

In mid or late summer, leaves on affected vines of dark-fruited cultivars such as Pinot noir may develop a chlorosis before they turn red, the wood at the base of canes may swell slightly, and the bark may split. This bark symptom is much less noticeable on *V. vinifera* cultivars than on hybrids.

The woody cylinder, cambium, and bark of many grape hybrids degenerate when infected. The hybrid LN-33 is especially sensitive to the disease (Table 2). In LN-33 the disease causes extensive cambium damage. Secondary phloem tissues in the bark proliferate, causing the bark to swell. At the same time the cambium and the outer xylem in the woody cylinder become totally deranged, and deep grooves appear. LN-33 vines die shortly after being inoculated. Grooving also develops in Rupestris Constantia, St. George, Harmony, C 1613, and Richter 110 vines following inoculation. Many *Vitis* species and selections are symptomless carriers of the disease. Grooving did not develop in 23 *Vitis* species, 38 hybrid rootstocks, and two *V. vinifera* cultivars when inoculated.

St. George, a selection from *V. rupestris* frequently used as a rootstock, shows severe symptoms when inoculated. Its bark becomes thick and appears rough. Removing a diseased St. George rootstock from the soil and cutting the trunk transversely reveals that the bark is thickened, the outer edge of the woody cylinder is convoluted, and the inner wood is pinkish (Plate 103).

Many *V. vinifera* cultivars that carry the disease agent do not show symptoms of disease until they are grafted onto American rootstocks. An incompatibility develops at the graft union, and the scion slowly dies. Scion cultivars grafted onto a diseased St. George rootstock may die, but the rootstock itself lives on as a wild, weedy vine. Before corky bark disease was understood, such vines were frequently used in the Napa Valley in California as a source of rootstock cuttings.

Causal Agent

Corky bark is a viruslike disease, but no virus has been isolated and proved to be its cause. A closterovirus has been observed in phloem tissues of affected vines.

Disease Cycle and Epidemiology

No vector for corky bark has been established, but the disease spreads rapidly in Aguascalientes State in Mexico, indicating that an aerial vector may be present in that area. The disease spreads only by propagation materials in other countries. Diseased and healthy vines have grown side-by-side for as long as 40 years in California vineyards.

Control

The use of propagation stocks from disease-free mother vines controls corky bark in all areas except central Mexico, which is the only region where spread by a natural vector has been reported.

Selected References

Beukman, E. F., and Goheen, A. C. 1970. Grape corky bark. Pages 207–209 in: Virus Diseases of Small Fruit and Grapevines (a Handbook). N. W. Frazier, ed. Division of Agricultural Sciences, University of California, Berkeley. 290 pp.

Goheen, A. C. 1981. Grape virus diseases. Pages 84–92 in: Grape Pest Management. D. L. Flaherty, F. L. Jensen, A. N. Kasimatis, H. Kido, and W. J. Moller, eds. Publ. 4105. Division of Agricultural Sciences, University of California, Berkeley. 312 pp.

Tzeng, H. L. 1984. Anatomical and tissue culture studies of corky-bark-, rupestris-stem-pitting-, and leafroll-affected grapevines. M.S. thesis. Department of Plant Pathology, University of California, Davis. 65 pp.

Tzeng, H. L., and Goheen, A. C. 1984. Electron microscopic studies on the corky bark and leafroll virus diseases of grapevines. (Abstr.) Phytopathology 74:1142.

(Prepared by A. C. Goheen)

Rupestris Stem Pitting

Rupestris stem pitting disease was first recognized in California in 1976 among selections that were being imported from Western Europe. A high percentage of such selections were affected. A similar high incidence was found in selections coming from Australia. Older California selections, which originated in Western Europe before 1950, were generally free of the disease.

Symptoms

The disease causes a slow decline in the growth of *V. vinifera* cultivars. After several years, affected vines are considerably smaller than healthy ones. The leaves on affected vines do not become yellow or red like those on vines affected by leafroll or corky bark. Rupestris stem pitting and leafroll have similar effects on fruit quality and vine yields.

The best indicator for the disease is St. George (Table 2), and chip bud grafting is the best method for inoculating indicator plants for indexing. A row of small pits develops below the point of inoculation on St. George (Plate 104). These pits occur on other American rootstocks also, but *V. rupestris* and its hybrids seem to develop the most evident pitting symptoms.

Rupestris stem pitting has frequently been confused with corky bark when inoculations have been made by machine grafts that are placed on the top of the indicator, because the pits that develop with this type of inoculation are distributed all around the woody cylinder. If the indicator is grafted onto the top of the candidate that is being tested, symptoms are rarely observed even when the candidate is infected. Rupestris stem pitting does not produce symptoms in LN-33, the indicator for corky bark.

Causal Agent

A closterovirus, which has not been isolated, has been observed in affected vines but has not been proved to have any causal relationship to the disease. It appears to be smaller than the closteroviruses associated with vines that are affected by leafroll and corky bark.

Disease Cycle and Epidemiology

The available evidence suggests that rupestris stem pitting spreads mainly by propagation. The incidence is high in selections of many grape cultivars coming from Western Europe and Australia. Indexing tests in California have shown that the disease is widespread in French-American hybrids growing in commercial vineyards in the northern and eastern United States and Canada. Close examination reveals pits in the woody cylinder of the trunks of some of these hybrids.

Control

Use of certified propagation materials controls rupestris stem pitting. All mother stocks of the Foundation Plant Materials Service of the University of California at Davis have been tested for this disease. Since 1981, only stocks that are free of this and other virus and viruslike diseases have been registered.

Selected Reference

Prudencio, S. 1985. Comparative effects of corky bark and rupestris stem pitting diseases on selected germplasm lines of grapes. M.S. thesis. Department of Plant Pathology, University of California, Davis. 36 pp.

(Prepared by A. C. Goheen)

Other Virus and Viruslike Diseases

Several additional viruses, especially nepoviruses, are found in grapevines in central Europe and elsewhere. Some of these cause minor diseases in grapevines and other host plants. Others infect vines but do not produce well-defined symptoms of disease. Several viruses that are found in grapevines seem to be contaminations because they show no known association with any disease symptom.

Minor Diseases Caused by Nepoviruses

Many nepoviruses closely related to but serologically distinct from grapevine fanleaf virus (GFLV) are spread by nematodes and are frequently found in grapevines. They can infect vines that have leafroll disease, and the effects of the dual infections are additive with respect to symptoms, growth, and yield.

In central Europe, investigators have isolated and identified seven nepoviruses from grapevines by serological methods, in addition to GFLV: Arabis mosaic virus, raspberry ringspot virus, tomato black ring virus, grapevine chrome mosaic virus, strawberry latent ringspot virus, artichoke Italian latent virus, and grapevine Bulgarian latent virus (= blueberry leaf mottle virus). Grapevine Bulgarian latent virus has been isolated from a single infected Concord grapevine in New York State.

Some of the nepoviruses seem to be of local importance. Their host range is wide. Some produce ringspot symptoms, but latent infections are also common. All are readily transmissible experimentally by sap inoculation to herbaceous test plants. Seed transmission is common in these herbaceous hosts.

Vectors, wherever they have been definitely established, are nematode species in the genera *Xiphinema* and *Longidorus*. The viruses are probably closely related.

To control the nepoviruses, grape clones that are free of them must be selected and tested. Virus-free clones are selected most effectively by serology. In recent years, enzyme-linked immunosorbent assay (ELISA) techniques have been widely used for this purpose.

Heat treatments have been used to eliminate the nepoviruses from diseased cuttings of several cultivars. Treatment of potted vines in greenhouses or growth chambers at 35–38° C for 84 days produced a 100% cure in Germany. Similar results have been achieved in California with GFLV after 60 days at 38° C. Such treatments eradicate nepoviruses from diseased tissues. Explant lines developed from the tissues that are no longer infected remain disease-free unless they are reinoculated by infective nematodes or budwood.

In central Europe, fallowing or soil fumigation is practiced to remove nematodes that might serve as virus vectors. A fallow period of up to 10 years may be necessary because of the persistence of nematodes on root fragments in the soil. Soil fumigation to control the complex of nematodes and soilborne viruses has had only limited success, in part because of the array of nematodes, viruses, and host plants involved in the nematode-virus problems of the region.

The best control for GFLV and other nepoviruses seems to be rootstocks that are nonhosts to either the nematodes or the viruses.

Contaminating Viruses

Certain aphidborne, fungusborne, or mechanically transmissible viruses, whose primary hosts are plants other than *Vitis* spp., are also found in grapevines in central Europe and elsewhere. Among these are alfalfa mosaic virus, broad bean wilt virus, sowbane mosaic virus, tobacco necrosis virus, petunia asteroid mosaic virus, Bratislava mosaic virus, and tobacco mosaic virus. These viruses are probably not primary pathogens of grapevines, and most are present in vines without causing damage.

Minor Viruslike Diseases

Several transmissible, viruslike diseases in grapes are poorly understood and have not been investigated in depth.

Fleck. Fleck disease is characterized in the indicator plant St. George by chlorotic, translucent spots (flecks) usually in the third- and fourth-order veins of young and medium-aged leaves (Plate 105). The flecks vary from 1 to 3 mm in length. The number of flecks on a leaf may vary from one to many. Leaves with numerous flecks are twisted and wrinkled. Fleck is graft-transmissible but is not transmitted through seed. It can be eliminated from affected vines by heat treatment but is more heat-resistant than the nepoviruses.

Fleck is distributed worldwide and occurs on many cultivars. Because fleck is often present in some of the best viticultural clones, the common presumption is that it does no harm, but this needs to be tested.

Fleck has been eliminated from registered mother vines in California because St. George indicator plants, used for indexing other diseases, also show clear symptoms of fleck. Until more is known about the disease, this is probably good insurance against a possible future problem.

Vein Mosaic. Vein mosaic disease produces more or less conspicuous symptoms on *V. vinifera* and other species. No causal agent has been established for it. It is reported from several European countries and from Australia.

The best indicator for vein mosaic is *V. riparia* 'Gloire de Montpellier'. On this indicator the typical symptom is a pale green mosaic affecting mostly the leaf tissue adjacent to the major veins, although symptoms are not necessarily restricted to such positions (Plate 106). In some cases the tissues tend to become necrotic. Although foliar symptoms are not severe, the disease adversely affects plant growth, reducing both root and shoot growth. In late summer, St. George shows reduced growth and general chlorosis of the basal leaves.

Enation. Enation disease produces enations on the underside of leaves (Plate 107). Enations are accompanied by various leaf malformations, including irregular shoot growth, cracking of the stems, and bud proliferation. In Italy the disease has been transmitted by grafting. The causal agent of enation is unknown, but the disease is often found on vines with severe GFLV infections.

Asteroid Mosaic. Asteroid mosaic disease produces small, star-shaped spots in leaves of affected vines. The spots seem to result when tissues around a veinlet degenerate. Tissue in the center of the spot may be necrotic. Affected leaves are asymmetrical. Symptoms become attenuated in summer. The disease also shows as irregular, blotchy veinbanding when graft-inoculated to St. George. Asteroid mosaic occurs now only in an isolated collection of diseased vines at Davis, California. It has not been observed in commercial vineyards in the past 25 years.

Yellow Speckle. Yellow speckle is a widely disseminated viruslike disease on several cultivars in Australia and California, but the expression of symptoms seems to require particular climatic conditions. The symptoms are ephemeral, consisting of a few to many spots on leaves (Plate 108). Symptoms are most evident at the end of summer, and they occasionally resemble those produced by the veinbanding syndrome of GFLV infection. The best indicator for the disease is the cultivar Esparte (syns. Mataro, Mourvedre). Heat treatments do not eliminate yellow speckle.

Shoot Necrosis. Shoot necrosis has been found only in southeastern Italy, where all vines of the cultivar Corniola appear to be affected. Symptoms are minute brownish spots and depressed striations, which develop early in the season at the base of very young shoots. These may expand and coalesce, causing extensive necrosis and splitting of the cortex. Shoots may die. Leaves are pale in the spring and the crop is apparently reduced, but the clusters are well shaped. The disease spreads with graft wood. No causal agent for shoot necrosis has yet been found, but it might be caused by a special strain of corky bark disease, which it closely resembles.

Selected References

Bercks, R. 1972. Die Serologie als Hilfsmittel bei der Erforschung und Bekämpfung von Rebenviren (unter Berücksichtigung von Erfahrungen bei anderen Kulturen). Weinberg Keller 19:481–487.

Bovey, R., Gärtel, W., Hewitt, W. B., Martelli, G. P., and Vuittenez, A. 1980. Virus and Viruslike Diseases of Grapevines. Payot, Lausanne; La Maison Rustique, Paris; and Verlag Eugen Ulmer, Stuttgart. 181 pp.

Frazier, N. W., ed. 1970. Virus Diseases of Small Fruits and Grapevines (a Handbook). Division of Agricultural Sciences, University of California, Berkeley. 190 pp.

Hewitt, W. B. 1979. On the origin and distribution of virus and viruslike diseases of the grapevine. Pages 3–5 in: Proc. 6th Meeting ICVG, Cordoba, Spain, 1976. Monografias INIA No. 18. Ministerio de Agricultura, Madrid.

Hewitt, W. B., Goheen, A. C., Corey, L., and Luhn, C. 1972. Grapevine fleck disease, latent in many varieties, is transmitted by graft inoculation. Ann. Phytopathol. (Hors Série):43–47.

Legin, R., and Vuittenez, A. 1973. Comparaison des symptômes et transmission par greffage d'une mosaique nervaire de *Vitis vinifera*, de la marbrure de *V. rupestris* et d'une affection nécrotique des nervures de l'hybride Rup.-Berl. 110 R. Riv. Patol. Veg. Ser. IV 9(Suppl.):57–63.

Prota, U., and Garau, R. 1979. Enations of grapevine in Sardinia. Pages 179–189 in: Proc. 6th Meeting ICVG, Cordoba, Spain, 1976. Monografias INIA No. 18. Ministerio de Agricultura, Madrid.

(Prepared by G. Stellmach and A. C. Goheen)

Nematode Parasites of Grapes

Nematode diseases of grape were first reported almost 100 years ago. These first reports were concerned exclusively with root-knot nematodes in the eastern United States, and very few reports were published until the mid-1950s. Since 1954, research has shown that grapes are afflicted with a wide array of nematode species, all of them root parasites. They occur worldwide wherever grapes are grown.

Nematode parasites are particularly insidious because symptoms of their attacks are obscure, generally comprising unthrifty, weakened vines. The evidence against some nematode species as serious pathogens is quite complete, but many others are known only because they have been detected in soil surveys. Much more work needs to be done to determine precisely the nematode-host interactions and to develop effective, economic control measures.

Root-Knot Nematodes

Bessey was the first to find and describe root-knot nematodes on species of *Vitis* in Florida in 1911. Since then, four species of *Meloidogyne* have been determined to be important pathogens of grape. They are found worldwide in every major region where grapes are cultivated.

Root-knot nematodes seldom kill vines; more often, plants decline in vigor and are more susceptible to stress. For example, severe leaf burn and even berry damage may occur on nematode-infested vines when hot weather closely follows sulfur treatments because the plants are unable to move enough water quickly enough to compensate. Yields decline to marginal levels, half or less than those of healthy vineyards in the same area. The steady decline in vine vigor can be slowed or offset in part by exercising special care in irrigation, by avoiding overcropping, and by controlling other diseases and pests that stress the plants.

Root-knot nematodes cause even greater damage in newly replanted vineyards. Young vines may fail to become established, and those that do are weak and do not grow enough to permit training onto stakes or trellises.

Symptoms

Root-knot nematodes, like all other nematode species pathogenic to grapevines, do not cause specific symptoms on aboveground parts of the vines. Lowered yields, poor growth, "off" color, and greater sensitivity to stress may accompany nematode attacks, but these conditions are often confused with water stress or nutrient deficiencies.

The distinctive response of grape roots to root-knot nematodes is the production of small swellings or galls on young feeder roots or secondary rootlets (Plate 109). Larger galls may result from multiple infections. When the galls are broken apart, the tiny, glistening, white bodies of mature females can often be detected with the aid of a hand lens. Second-stage juveniles and males can only be found after sieving soil on screens and examining the residues under a dissecting microscope. Usually the root systems of vines attacked by root-knot nematodes are severely limited and many feeder roots are dead.

One root-knot species, *M. nataliei* Golden, Rose & Bird, has been reported to parasitize grape without gall production on the roots. It is known only from one vineyard in Michigan but must be considered in future surveys of vineyard problems.

Causal Organisms

Root-knot nematodes are members of the genus *Meloidogyne* Goeldi. The four species important to grape production are *M. incognita* (Kofoid & White) Chitwood, *M. javanica* (Treub) Chitwood, *M. arenaria* (Neal) Chitwood, and *M. hapla* Chitwood.

Life Cycle and Epidemiology

Root-knot females deposit eggs outside their body, mostly in a gelatinous matrix holding up to 1,500 eggs. The matrix is usually outside the root but may be inside and completely surrounded by root tissue. The young develop into an elongate shape, molt once, and emerge from the egg as second-stage juveniles. These juveniles are the migratory stage. They locate new feeding sites by penetrating the root cortex and complete their life cycle as sedentary endoparasites. The plant response to nematode feeding is the production of multinucleate "giant cells."

The juvenile, without further feeding, quickly sheds its cuticle three more times to become an adult female, pyriform in shape. The cycle from egg to egg takes about 25 days at 27° C, and several generations a year are possible. The nematodes overwinter principally as developing juveniles inside eggs in the matrix.

Root-knot nematodes are mostly parthenogenetic. Males are rare or lacking in the soil but may appear when roots become overcrowded or in response to other conditions stressful to nematodes. The male cycle is similar to that of females up to the third stage, when the developing young elongate into slender fourth-stage juveniles, then into adults.

Root-knot nematodes are spread within a field or to new fields usually by infected rootings or by cultivation practices.

Control

Exclusion. Once established, root-knot nematode infestations of soils are permanent. However, the parasite is not universally distributed and should be kept out of any fields known to be free of it. Some states and countries have regulatory agencies empowered to control the introduction and spread of pests and diseases, including nematodes. Growers should survey their lands to determine which fields have nematodes and should take every precaution to keep nematodes out of clean fields. Because the most common and effective way nematodes are brought into new fields is in infected rootings, growers should use certified nematode-free rootings for planting stock.

Side-Dressing of Established Vineyards. Chemicals that are effective as soil treatments for nematode control are generally phytotoxic. An exception is 1,2-dibromo-3-chloropropane (DBCP), which was discovered about 1950. DBCP functioned as a fumigant and worked best when applied as an emulsion in irrigation water. It was selectively nematicidal and persistent, which accounted for most of its effectiveness. However, the persistence also enabled spread of DBCP into groundwaters, and it was removed from the market in California in 1977.

Several nonfumigant, systemic-type nematicides have been tested over a period of several decades. These chemicals, granular or water-soluble formulations of organophosphates or carbamates, do not diffuse through soil but must be mixed with soil or water and eventually carried by water. Disadvantages of these materials include restricted movement of water in clay loam and clay-type soils, residues that may occur in fruit, and the high cost of the treatments.

Replant Treatments. Vineyards that decline in productivity and do not respond to side-dressing treatments must be considered for replanting. For almost two decades, fumigation with 1,3-dichloropropene (1,3-D) or methyl bromide has been an economical and successful soil treatment for new vineyards established in nematode-infested soils, when recommendations have been followed carefully. Vines should be removed with a cutter below the crown, not pulled out by chains, which usually break off the trunk at ground level. Once a vineyard is removed, a fallow or host-free period of at least one year, preferably up to four years, should be allowed before replanting. Soil should be ripped deeply (0.8–1.5 m) at 1-m spacing in two perpendicular directions and fumigated. The new planting stock should be certified nematode-free.

The recommended dosage of 1,3-D is about 1,400 L/ha applied 0.5–1.0 m deep on 1-m spacing. Methyl bromide is used at 330–400 kg/ha under continuous polyethylene cover or 350–500 kg/ha without cover (the lower rate in sandy soils) applied 0.6–0.8 m deep on 1.7-m spacing. Some treatments have given good results at 550–660 kg/ha without the cover. Planting must be delayed 10–14 days after treatment with methyl bromide and much longer (three to four months) with 1,3-D at these dosages.

Vineyards planted in fumigated soils have consistently shown a high percentage of established vines and uniformly vigorous growth. However, no treatments have ever eradicated nematodes from infested soils. Careful surveillance is important to detect nematode population buildup as early as possible. Side-dressing treatments may be needed to keep the population levels down.

Resistant Rootstocks. Until recently the search for nematode resistance in rootstocks was directed only at root-knot nematodes. Dog Ridge and Salt Creek (also known as Ramsey) have high resistance, almost immunity, to root-knot nematodes. Both have excess vigor in most soils except very coarse sands, are difficult to root and establish, and show excessive suckering. Nevertheless, they do bud and graft readily. Unfortunately, both rootstocks are susceptible to *Xiphinema index,* the nematode vector of grapevine fanleaf virus and a pathogen itself.

Freedom and Harmony rootstocks are resistant, but not immune, to root-knot nematodes and phylloxera. These rootstocks are becoming increasingly popular in California but need further research because some selections of Harmony are reported to support large numbers of *X. index,* and there is also evidence that high populations of *M. arenaria* may build up on Harmony rootstock. Because a wide variety of nematodes attack grapes, multiple resistance is needed but so far is not available in one rootstock.

Selected References

Lider, L. A. 1959. Nematode resistant rootstocks for California vineyards. Calif. Agric. Exp. Stn. Leafl. 114.

Raski, D. J., Hart, W. H., and Kasimatis, A. N. 1973. Nematodes and their control in vineyards. Calif. Agric. Exp. Stn. Circ. 533 (revised). 20 pp.

Raski, D. J., Jones, N. O., Kissler, J. J., and Luvisi, D. A. 1976. Soil fumigation: One way to cleanse nematode-infested vineyard lands. Calif. Agric. 30:4–6.

Raski, D. J., Jones, N. O., Hafez, S. L., Kissler, J. J., and Luvisi, D. A. 1981. Systemic nematicides tested as alternatives to DBCP. Calif. Agric. 35:11–12.

Sauer, M. R. 1962. Distribution of plant-parasitic nematodes in irrigated vineyards at Merbein and Robinvale. Aust. J. Exp. Agric. Anim. Husb. 2:8–11.

Tyler, J. 1933. Development of the root-knot nematode as affected by temperature. Hilgardia 7:389–415.

Tyler, J. 1944. The root-knot nematode. Calif. Agric. Exp. Stn. Circ. 330 (revised July 1944). 30 pp.

(Prepared by D. J. Raski)

Dagger and Needle Nematodes

One or more of 10 species of dagger nematodes, *Xiphinema* spp., are present in all major grape-growing areas of the world. *X. index* is the most thoroughly documented of these species and is a devastating pathogen of grapes. It is worldwide in distribution. Affected vines decline markedly, producing fewer shoots until ultimately vines are totally unproductive. *X. index* is also a vector of grapevine fanleaf virus (GFLV), and infestations of viruliferous *X. index* have even more devastating effects on vineyards, which quickly become uneconomical. *X. americanum* is also widespread (though not in European vineyards), but little is known of its pathogenicity. The other eight species have received little attention, and many are known only from survey data.

At least seven species of needle nematodes, *Longidorus* spp., are reported from vineyard soils. Two of the species have been described as causing necrosis and malformations of grape roots. The remaining species have not been studied and are known only as part of recorded collections.

Once decline caused by these nematodes begins, little can be done to restore the affected vines' vigor and productivity. Most infestations cause slow, gradual decline, seldom killing the vines outright.

Symptoms

The root system of affected vines exhibits many dead feeder roots, which can result in a kind of "witches'-broom" effect. Some distinctive plant responses also accompany dagger nematode attacks. Most feeding takes place near root tips, and root growth stops soon after feeding begins. Proliferation of cells by hyperplasia and enlargement by hypertrophy often cause a bending accompanied by slight swelling (Plate 110). Multiple prolonged attacks can result in darkened, necrotic spots that spread over the entire root tip. These plant responses are characteristic for *X. index* and *X. diversicaudatum* and are less so for *X. americanum.*

Causal Organisms

The species of dagger nematodes that have been associated with grape are *X. americanum* Cobb, *X. index* Thorne & Allen, *X. italiae* Meyl, *X. diversicaudatum* (Micoletzky) Thorne, *X. mediterraneum* Martelli & Lamberti, *X. pachtaicum* (Tulagonov) Kirjanova, *X. brevicolle* Lordello & Da Costa, *X. algeriense* Luc & Kostadinov, *X. vuittenezi* Luc, Lima, Weischer & Flegg, and *X. turcicum* Luc & Dalmasso.

The species of needle nematodes that have been associated with grape are *L. attenuatus* Hooper, *L. elongatus* (de Man) Thorne & Swanger, *L. sylphus* Thorne, *L. diadecturus* Eveleigh & Allen, *L. iranicus* Sturhan & Barooti, *L. macrosoma* Hooper, and *L. protae* Lamberti & Bleve-Zacheo.

Life Cycle and Epidemiology

Dagger and needle nematodes have four juvenile stages and adults with sexes separate (Fig. 29). Their life cycle differs from that of root-knot nematodes in that they hatch from eggs as first-stage juveniles and molt four times in the soil to become adults. The young resemble the adults, and all remain vermiform, without swellings or modifications such as cysts.

Dagger and needle nematodes are strictly ectoparasites. They feed by means of a very long stylet that they use to penetrate the vascular system of roots. They do not produce a matrix or special covering for their eggs. Each stage must feed before it can molt and continue growth. Reproduction in some species is mostly parthenogenetic, and males often are rare or not known. In other species, males are present in about the same numbers as females.

The cycle from egg to adult female in *X. index* is 22–27 days in California but is reported to be seven to nine months for a complete generation in Israel. Reasons for this difference are not known.

These nematodes spread to new fields generally by means of infected rootings or as a result of cultural practices, sometimes in contaminated irrigation water.

Control

As is typical of all nematodes, field infestations of dagger and needle nematodes are permanent. Techniques for excluding new infestations are the same as for root-knot nematodes.

With regard to side-dressing of established vines, the fumigant 1,2-dibromo-3-chloropropane (DBCP), before it was discontinued, had some of its most remarkable responses in grapes where *X. index* was the principal pathogen but was free of GFLV. One treatment controlled the dagger nematode for several years, and growth and yield responses were dramatic. Newer, nonfumigant nematicides have not yet given as effective control.

Replant treatments with 1,3-dichloropropene or methyl bromide as described for root-knot nematodes have given good control and resulted in vigorous, productive vineyards. However, buildup of surviving nematodes and, finally, spread of GFLV when present inevitably follow. The economic life expectancy of such a replanted vineyard is 12–20 years.

Few rootstock selections have shown satisfactory resistance to *X. index,* and no rootstocks have been tested for resistance to other species of dagger and needle nematodes. Two new rootstocks selected from crosses of *V. rotundifolia* and *V. vinifera* have shown significant promise for control of both *X. index* and GFLV.

Selected References

Cohn, E. 1970. Observations on the feeding and symptomatology of *Xiphinema* and *Longidorus* on selected host roots. J. Nematol. 2:167–173.

Cohn, E., and Mordechai, M. 1969. Investigations on the life cycles and host preference of some species of *Xiphinema* and *Longidorus* under controlled conditions. Nematologica 15:295–302.

Cotton, J. 1975. Virus vector species of *Xiphinema* and *Longidorus* in relation to certification schemes for fruit and hops in England. Pages 283–285 in: Nematode Vectors of Plant Viruses. F. Lamberti, C. E. Taylor, and J. W. Seinhorst, eds. Plenum, New York. 460 pp.

Das, S., and Raski, D. J. 1968. Vector-efficiency in *Xiphinema index* in the transmission of grapevine fanleaf virus. Nematologica 14:55–62.

Fisher, J. M., and Raski, D. J. 1967. Feeding of *Xiphinema index* and *X. diversicaudatum.* Proc. Helminthol. Soc. Wash. 34:68–72.

Hewitt, W. B., Raski, D. J., and Goheen, A. C. 1958. Nematode vector of soil-borne fanleaf virus of grapevines. Phytopathology 48:586–595.

Martelli, G. P. 1978. Nematode-borne viruses of grapevine, their epidemiology and control. Nematol. Mediterr. 6:1–27.

Pinochet, J., Raski, D. J., and Goheen, A. C. 1976. Effects of *Pratylenchus vulnus* and *Xiphinema index* singly and combined in vine growth of *Vitis vinifera.* J. Nematol. 8:330–335.

Raski, D. J., and Schmitt, R. V. 1972. Progress in control of nematodes of soil fumigation in nematode-fanleaf infected vineyards. Plant Dis. Rep. 56:1031–1035.

(Prepared by D. J. Raski)

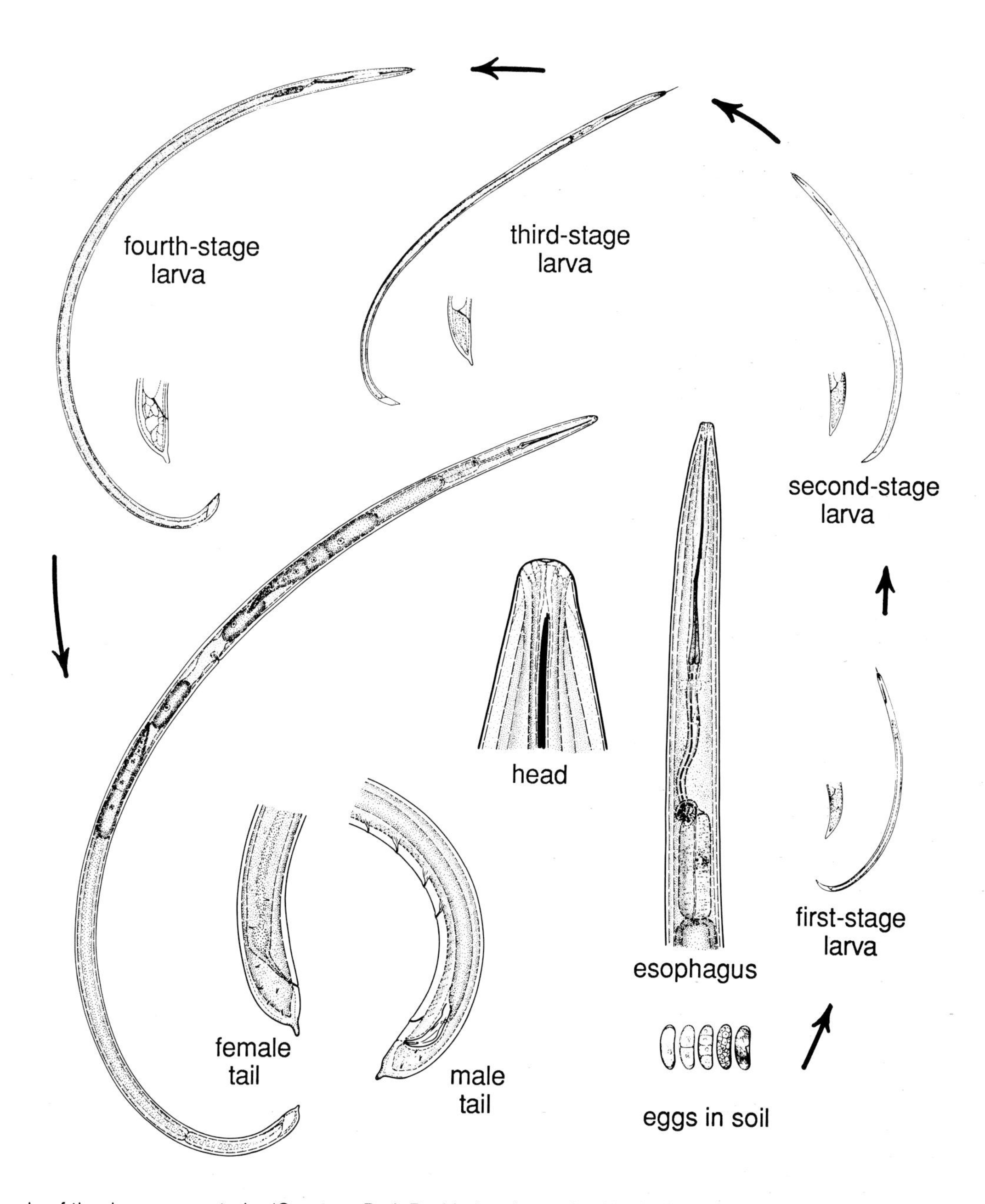

Fig. 29. Life cycle of the dagger nematode. (Courtesy D. J. Raski; drawing revised by R. Sticht)

Lesion Nematodes

The earliest records of lesion nematodes attacking grapes were published after 1950. Five species of *Pratylenchus* are known from vineyard surveys. The most important species, *P. vulnus,* is found in localized spots throughout the San Joaquin Valley in California and is also fairly widespread in Australia on grapes. *P. pratensis* is found in Tashkent in the Soviet Union. The other three species have limited distribution in Australia and California.

Damage caused by lesion nematodes is more severe than that caused by root-knot nematodes, and once decline sets in, vines do not respond to cultural practices aimed at alleviating injury.

Symptoms

Root symptoms of lesion nematode attack are most often poor root development and many dead feeder roots. Occasionally rootlets are killed, as soon as or shortly after they are produced, ultimately resulting in a matted or "witches'-broom" appearance. This symptom is not necessarily specific for nematode injury, and there are no specific symptoms in grapes that are diagnostic for lesion nematode damage.

Causal Organisms

Lesion nematodes are members of the genus *Pratylenchus* Filipjev. The five species associated with grape are *P. vulnus* Allen & Jensen, *P. brachyurus* (Godfrey) Filipjev & Schuurmans Stekhoven, *P. scribneri* Steiner, *P. neglectus* (Rensch) Filipjev & Schuurmans Stekhoven, and *P. pratensis* (de Man) Filipjev. *P. vulnus* is by far the most widespread, but the other four species are potentially serious pathogens of grape and must be considered in any nematode surveys and in research on control by resistant rootstocks.

The genus *Zygotylenchus* Siddiqi belongs to the same nematode family as *Pratylenchus.* A few records of *Zygotylenchus* spp. have been published in surveys of vineyard soils, but nothing more is known about their abundance, general incidence, or significance in grape production.

Life Cycle and Epidemiology

Lesion nematodes deposit eggs singly in the soil or in root tissues. They are migratory and endoparasitic in life habits. The young develop and elongate, molt once, and emerge from eggs as second-stage larvae. These larvae penetrate host roots and move through the cortex, penetrating, feeding on, and killing cells. In some hosts certain polyphenolic compounds are oxidized, causing necrosis and lesion formation. Males are usually common, and sexual reproduction occurs.

Control

Exclusion practices, side-dressing of established vineyards, and replant treatments are all parallel to those described for root-knot nematodes. Because *P. vulnus* can attack rootstocks that have been developed for their resistance to root-knot species, all candidate rootstocks should be tested for their resistance to lesion nematodes to ensure their performance in the presence of these nematodes.

Selected References

Allen, M. W., and Jensen, H. J. 1951. *Pratylenchus vulnus,* new species (Nematoda: Pratylenchinae), a parasite of trees and vines in California. Proc. Helminthol. Soc. Wash. 18:47–50.

Pinochet, J., and Raski, D. J. 1977. Observations on the host-parasite relationship of *Pratylenchus vulnus* on grapevine, *Vitis vinifera.* J. Nematol. 9:87–88.

Pinochet, J., Raski, D. J., and Goheen, A. C. 1976. Effects of *Pratylenchus vulnus* and *Xiphinema index* singly and combined in vine growth of *Vitis vinifera.* J. Nematol. 8:330–333.

Sauer, M. R. 1962. Distribution of plant-parasitic nematodes in irrigated vineyards at Merbein and Robinvale. Aust. J. Exp. Agric. Anim. Husb. 2:8–11.

Sher, S. A., and Allen, M. W. 1953. Revision of the genus *Pratylenchus* (Nematoda: Tylenchidae). Univ. Calif. Berkeley Publ. Zool. 57:441–470.

(Prepared by D. J. Raski)

Citrus Nematode

The citrus nematode was first found on citrus roots in 1913 and is now known to be worldwide in distribution on that crop. The first record of citrus nematode on grape was in California in 1956, and in that same year it was also found in vineyards in Australia. Since then it has been reported from vineyards in India, Egypt, and The Philippines. Future surveys undoubtedly will reveal more occurrences of this nematode, especially where grapes follow citrus.

The citrus nematode is considered one of the most pathogenic nematode species on grape. Vigor is remarkably reduced, and affected plants do not have resilience to withstand stressful conditions. Yields gradually and inevitably decline, and vineyards become uneconomical.

Symptoms

No specific symptoms diagnostic for the citrus nematode are produced on aboveground parts of grape. The main effect on roots is the death of feeder roots, although some rootlets may support large populations. The nematodes produce a profuse, gelatinous matrix to which soil particles cling, giving them a "dirty" appearance (Plate 111). No necrotic spots or malformations are evident.

Causal Organism

The citrus nematode is *Tylenchulus semipenetrans* Cobb. The genus *Tylenchulus* Cobb has only one other species, *T. furcus* van den Berg & Spaull.

Life Cycle and Epidemiology

The female citrus nematode is semiendoparasitic and sessile. It remains attached to the host root at its head end. The head is located in an empty cell, and the female feeds on surrounding "nurse" cells.

The life cycle requires four to eight weeks to complete. The young emerge from eggs as second-stage larvae. Males undergo three more molts quickly without feeding, having a degenerate stylet and esophageal system. They are found only in the soil and are not necessary for reproduction. The females feed in the cortical tissues and also molt three more times before becoming adults embedded in the root for the rest of their life cycle.

Control

Exclusion tactics are the same as for root-knot nematodes. Side-dressing of established vineyards has some potential for success because the citrus nematode is semiendoparasitic and partly exposed to chemical treatments. However, the gelatinous matrices afford some protection to the females and eggs. So far, no specific materials effective against the citrus nematode on grape are economically feasible. Replant treatments for new vineyards are similar to those for root-knot nematodes. No fruiting cultivars or rootstocks of *Vitis* have been tested specifically for citrus nematode resistance.

Selected References

Cobb, N. A. 1913. Notes on *Mononchus* and *Tylenchulus.* J. Wash. Acad. Sci. 3:287–288.

Cobb, N. A. 1914. Citrus-root nematode. J. Agric. Res. 2:217–230.

Raski, D. J., Sher, S. A., and Jensen, F. N. 1956. New host records of the citrus nematode in California. Plant Dis. Rep. 40:1047–1048.

Sauer, M. R. 1962. Distribution of plant-parasitic nematodes in irrigated vineyards at Merbein and Robinvale. Aust. J. Exp. Agric. Anim. Husb. 2:8–11.

Siddiqi, M. R. 1974. *Tylenchulus semipenetrans*. Descriptions of Plant-Parasitic Nematodes, Set 3, No. 34. Commonwealth Institute of Helminthology, St. Albans, Hertfordshire, England. 4 pp.

Van Gundy, S. D. 1958. The life history of the citrus nematode *Tylenchulus semipenetrans* Cobb. Nematologica 3:283–294.

(Prepared by D. J. Raski)

Miscellaneous Ectoparasitic Nematodes

Among the ectoparasitic nematode species that have been found in vineyard soils are ring nematodes, *Criconemella xenoplax* (Raski) Luc & Raski; pin nematodes, *Paratylenchus hamatus* Thorne and *P. neoamblycephalus* Geraert; reniform nematodes, *Rotylenchulus* spp.; spiral nematodes, *Helicotylenchus* spp.; lance nematodes, *Hoplolaimus* spp. and some species of the related genus *Rotylenchus*; the stubby-root nematode, *Paratrichodorus christiei* (Allen) Siddiqi; and stunt nematodes, *Tylenchorhynchus* (syn. *Telotylenchus*) spp.

Almost no information on these parasites of grape was available until the mid-1950s. Since then, information has developed slowly, and in recent years nematode surveys in numerous countries have revealed a wide array of ectoparasitic species in vineyard soils. Their importance to grape production remains to be assessed.

Selected References

Pinochet, J., Raski, D. J., and Jones, N. O. 1976. Effect of *Helicotylenchus pseudorobutus* on Thompson Seedless grape. Plant Dis. Rep. 60:528–529.

Raski, D. J. 1955. Additional observations on the nematodes attacking grapevines and their control. Am. J. Enol. Vitic. 6:29–31.

Raski, D. J., and Radewald, J. D. 1958. Reproduction and symptomatology of certain ectoparasitic nematodes on roots of Thompson Seedless grape. Plant Dis. Rep. 42:941–943.

(Prepared by D. J. Raski)

Part II. Mites and Insects That Cause Diseaselike Symptoms in Grapes

Mites

Spider Mites

Spider mites are common pests of grapevines. Two species are considered important pests in California: the Pacific spider mite, *Tetranychus pacificus* McGregor, and the Willamette spider mite, *Eotetranychus willamettei* (Ewing). In the eastern United States, the European red mite, *Panonychus ulmi* (Koch), is the principal spider mite pest. In Europe, *P. ulmi, E. carpini* Oudemans, *T. urticae* Koch (the twospotted spider mite), and *T. mcdanieli* McGregor are grape pests. *Oligonychus vitis* Zaher & Shehata is reported as a serious pest of grapes in Egypt and Chile. A number of spider mites, including *P. ulmi, E. pruni* (Oud.), *T. urticae, T. turkestani* Ugarov & Nikolski, and *Bryobia praetiosa* Koch, are reported to infest vineyards in the Soviet Union.

Willamette and Pacific spider mites. In vineyards in the San Joaquin Valley in California, feeding by small colonies of Willamette and Pacific spider mites produces yellow spots on the upper leaf surfaces. Yellowing of foliage is characteristic of high densities of Willamette spider mites (Plate 112). A bronze discoloration, heavy webbing, and cupping of the foliage are characteristic symptoms of high Pacific spider mite populations, particularly on foliage on top of vines exposed to the sun (Plate 113). Later, heavily infested vines may become defoliated as leaves turn brown and dry.

Pacific spider mite injury in vineyards is spotty and tends to occur on weaker vines or in water-stressed areas. Both Pacific and Willamette spider mite populations are favored by dusty vineyard conditions. Yellowing of vines caused by nitrogen deficiency is sometimes mistaken for Willamette spider mite injury. On dark cultivars such as Cabernet Sauvignon in the cooler coastal areas of California, Willamette spider mite feeding causes a reddish coloration of the foliage, which is delimited by veinlets and occurs primarily on the north side or shady areas of the vines on east-west rows. This injury may be mistaken for leafroll (Plate 114).

Willamette spider mites may cause amber coloring of Thompson Seedless table grapes if clusters are exposed to direct sunlight. Severe injury by Willamette spider mites has been reported on Zinfandel in northern California. Large numbers of overwintering Willamette or Pacific spider mites may cause necrotic areas on foliage in early spring.

Twospotted spider mite. In California, twospotted spider mite injury is similar to that of the Pacific spider mite on young shoot growth. Infestations are often related to movement from nearby heavily infested fruit orchards. The mites reproduce rapidly and cause severe injury to new growth on vines, but reproduction appears to cease as foliage matures.

In Europe, the twospotted spider mite is an important pest of grapevines in regions with dry summers. It is particularly serious in Spain. The first symptoms of attack are chlorotic spots caused by feeding on lower leaf surfaces where dense colonies thrive. Later, defoliation is likely to affect berry maturation and quality. Twospotted mites also attack berries, producing dark spots on the skin.

European red mite. In the eastern United States, injury by European red mites is recognized as a fine speckling that appears at the base of leaves. The speckling develops into a bronze coloration (Plate 115) that soon covers the entire leaf with symptoms resembling leafroll. Continued feeding by the pest may cause leaves to turn brown and fall. Oxidant stipple is sometimes confused with European red mite bronzing. However, leaves with oxidant stipple are usually darker brown than those bronzed by mites.

In France, European red mites may seriously injure new growth at budbreak. The margins of young leaves appear brown as if damaged by a mild frost. The injured leaves remain small and deformed and often appear yellow-mottled. When attack is heavy, injured leaves shrivel, desiccate, and fall. In general, emerging shoots from buds at the base of canes or spurs are more severely injured because they are close to older wood where mites overwinter and they grow more slowly than shoots from terminal buds. During the period of rapid vine growth, injury becomes less visible, and mites may not be noticed again until the end of summer, when foliage is bronzed on dark cultivars or yellowed on white cultivars.

E. carpini. Commonly called the yellow vine mite, *E. carpini* occurs on grapevines in Mediterranean viticultural areas, especially in France and Italy. The injury it causes is similar to that caused by the Willamette spider mite in California. Overwintering females attack the young shoots during budbreak, causing small necrotic spots.

T. mcdanieli. Early-season injury caused by *T. mcdanieli* in France is similar to that caused by *T. pacificus* in California. Young tissue yellows, along with some diffused grayness and necrosis. Young leaves show cupping, with some drying of the raised margins. Summer injury resembles leafroll. Pinot noir is highly susceptible to attack by *T. mcdanieli.*

O. vitis. In desert areas in northern Chile, *O. vitis,* which characteristically infests upper leaf surfaces, causes a heavy bronzing in late summer and may cause early defoliation of the cultivars Emperor, Thompson Seedless (Sultanina), and Muscat of Alexandria. In Egypt the cultivar Romi-red is commonly infested.

Grape Erineum Mite

The grape erineum mite, *Colomerus vitis* (Pagenstecher), is an eriophyid mite that attacks various species of grapes and is probably widely distributed geographically. Three strains of the mite have been recognized by the characteristic injury each strain causes: the erineum strain, the bud mite strain, and the leaf curl strain. The strains are morphologically identical.

Adults are whitish and wormlike, approximately 0.2 mm long and less than 0.05 mm wide.

Erineum Strain. The erineum strain feeds on leaves and causes patches of concave, felty galls called erinea on the lower leaf surface, followed by blisterlike swellings on the upper surface. The galls are whitish at first, then yellow, and finally reddish brown (Plate 116). In the white stage, the galls can be confused with sporulation of the downy mildew fungus. A leaf may have as many as 50 erinea. These leaves may drop in the fall somewhat earlier than noninfested leaves.

Bud Mite Strain. The bud mite strain lives within grape buds and does not produce erinea on leaves. Bud mites are generally confined to feeding on the outer bud scales, but they may penetrate and feed on embryonic tissues of the shoot primordia. Common symptoms (Plate 117) include short basal internodes, scarred epidermis of new shoots, flattened shoots, dead terminal buds on new shoots, witches'-broom growth of new shoots, zigzagged shoots, and dead overwintering buds. Leaves are usually stunted and wrinkled, and the veins are prominent and drawn together. Flower clusters may also drop prematurely because of injury to immature leaf and flower buds. Bud mite is only rarely encountered. It has been confused with boron deficiency in early spring.

Leaf Curl Strain. Symptoms of the leaf curl strain appear in summer and include downward curling or rolling of leaves (Plate 118). These symptoms are sometimes confused with the leaf cupping symptoms of boron toxicity. The rolling ranges from slight to severe curling where the leaf tends to roll into a crude ball. Overall growth of the shoot may be stunted, and scarring is sometimes noted on the shoot. Lateral shoot growth may also be present.

Grape Rust Mite

The grape rust mite, *Calepitrimerus vitis* (Nalepa), is an eriophyid mite. These light amber mites are 0.15 mm long and appear wormlike; the front end is broader than the rear. They move slowly on the leaf surface. The mite has been reported as a pest in California, Portugal, France, and the Soviet Union.

Heavy infestations prevent vines from growing normally early in the season. Buds are killed, internodes are shortened, foliage becomes dense, and fruit production is reduced. Clusters are damaged when flowers are injured. Feeding on the surface of leaves of white grape cultivars causes a yellowing that closely resembles the appearance of leaves slightly injured by spider mites. On dark grape cultivars, injured leaves become brilliant red, as they do in grape diseases such as leafroll.

False Spider Mites

False spider mites are less than 0.5 mm long and resemble very small, flat, red spider mites. *Brevipalpus chilensis* Baker is a serious pest in Chile. In 1984, damaging populations of *B. lewisi* McGregor were observed for the first time in California vineyards. *B. lewisi* is responsible for "bunch mite" symptoms of grapes in Australia and is also reported as a pest in Greece and the Soviet Union. Another false spider mite species, *Tenuipalpus granati* Sayed, is reported as a pest of grapevines in Egypt.

In Chile, *B. chilensis* injures shoots and foliage of dark cultivars. Cultivars grafted on American rootstocks are attacked to a lesser extent. During budbreak, overwintering *B. chilensis* females induce necrosis of leaves and shoots, which acquire an intense brown color similar to the damage produced by early frost. As populations increase and spread, leaves become discolored and their margins fold down. The leaves first appear greenish copper, then dark reddish, and finally grayish brown. Under heavy attack, newly developed leaves are smaller than normal and fruit production is reduced.

Injury may be especially severe in vineyards of Cardinal, Emperor, and Ribier table grapes. Complete dehydration and blackening of new shoot growth are not uncommon. Later, the mites distribute on the rachis, branches, and pedicels of clusters, where dehydration results in darkened grooves that unite to form large lesions (Plate 119) similar to several grape diseases such as Phomopsis cane and leaf spot and stem necrosis. The stems and berries may completely dehydrate. On the foliage, *B. chilensis* causes loss of green color and a concentration of anthocyanins on the lamina of dark cultivars, and affected parts appear reddish. Cultivars such as Moscatel de Austria tolerate high populations per leaf; the reddish mites concentrate on both sides of the main vein, especially in the vein angles.

Selected References

Buchannan, G. A., Bengston, M., and Exley, E. M. 1980. Population growth of *Brevipalpus lewisi* McGregor (Acarina: Tenuipalpidae) on grapevines. Aust. J. Agric. Res. 31:957–965.

Carmona, M. N. 1978. *Calepitrimerus vitis* (Nalepa), responsável pela "acariose da videira" I. Notas sobre a morfologia, biologia e sintomatologia. Agron. Lusit. 39:29–56.

Flaherty, D. L., Hoy, M. A., Lynn, C. D., and Peacock, W. L. 1981. Spider mites. Pages 111–125 in: Grape Pest Management. D. L. Flaherty, F. L. Jensen, A. N. Kasimatis, H. Kido, and W. J. Moller, eds. Publ. 4105. Division of Agricultural Sciences, University of California, Berkeley. 312 pp.

Gonzalez, R. H. 1983. Manejo de plagas de la vid. Cienc. Agric. No. 13. Departamento de Sanidad Vegetal, Universidad de Chile, Santiago. 115 pp.

Jeppson, L. T., Keifer, H. H., and Baker, E. W. 1975. Mites Injurious to Economic Plants. University of California Press, Berkeley. 614 pp.

Jubb, G. L., Jr. 1976. Vineyard insect pests: The European red mite. Eastern Grape Grower 2:14–15.

Keifer, H. H., Baker, E. W., Kono, T., Delfinado, M., and Styer, W. E. 1982. An Illustrated Guide to Plant Abnormalities Caused by Eriophyid Mites in North America. U.S. Dep. Agric. Agric. Handb. 573. 178 pp.

Kido, H. 1981. Grape erineum mite. Pages 217–220 in: Grape Pest Management. D. L. Flaherty, F. L. Jensen, A. N. Kasimatis, H. Kido, and W. J. Moller, eds. Publ. 4105. Division of Agricultural Sciences, University of California, Berkeley. 312 pp.

Schruft, G. A. 1986. Grape. Pages 354–366 in: Spider Mites, Their Biology, Natural Enemies and Control. W. Helle and M. W. Sabelis, eds. Elsevier, Amsterdam. 458 pp.

(Prepared by D. L. Flaherty and L. T. Wilson)

Thrips

Two species of thrips cause concern in California vineyards, mostly because they injure clusters of table grapes: western flower thrips, *Frankliniella occidentalis* (Pergande), and European grape thrips, *Drepanothrips reuteri* Uzel. In the eastern United States, *F. tritici* (Fitch), sometimes called the eastern flower thrips, and *D. reuteri* are considered potential pests. In Chile *F. cestrum* Moulton, also called flower thrips, and *D. reuteri* are important pests of table grapes. In Switzerland, Italy, France, Spain, Greece, Algeria, and Egypt, *D. reuteri* is primarily a pest of new shoot growth in spring and foliage in summer.

Western flower thrips injury on grapevines in California consists of halo spotting, which makes the fruit of certain white cultivars such as Calmeria, Almeria, and Italia unsightly and unmarketable; starfish scarring of Thompson Seedless (Sultanina) berries (Plate 120), which renders the clusters unsalable; and shoot stunting and foliage injury. The European grape thrips is mainly responsible for summer foliage damage, although it sometimes causes troublesome fruit scarring and shoot stunting. White Malaga grapes are particularly susceptible to fruit scarring by the European grape thrips. Exposed fruit on vines with a light canopy of foliage are most susceptible to injury.

Halo spots are caused by oviposition of the western flower thrips. A small, dark scar is produced at the site of the oviposition puncture. The tissue in a roughly circular area around this puncture becomes whitish. As the berries grow, these spots may result in growth cracks that allow the entry of rot organisms.

Starfish scarring is caused by nymphs of the western flower thrips feeding under caps (calyptras) and injuring the surface tissue of young berries (Plate 120). It resembles injury caused by pesticides. The nymphs cause scarring only when the flower parts fail to shed normally. Usually the caps adhere at the stylar end of the berry to produce the starfish scar. Larger berries resulting from the use of gibberellin also show larger, more visible starfish scarring.

European grape thrips injure fruit when the berries are 3 mm in diameter. Injury is attributed to the feeding of both adults and, to a greater extent, nymphs, and is found scattered within vineyards. White cultivars are most sensitive. In severe cases of scarring (Plate 121), White Malaga berries split and crack as they grow. This injury may resemble phytotoxicity caused by pesticides.

Western flower thrips and European grape thrips may cause shoot injury in early spring. European grape thrips injury to foliage is often severe during summer (Plate 122). Leaves are bronzed and fail to grow to normal size; internodes are shortened and scarred. Populations often have decreased by the time damage becomes evident. The symptoms might be confused with Eutypa dieback or boron deficiency.

Selected References

Gonzalez, R. H. 1983. Manejo de plagas de la vid. Cienc. Agric. No. 13. Departamento de Sanidad Vegetal, Universidad de Chile, Santiago. 115 pp.

Jensen, F. L., Flaherty, D. L., and Luvisi, D. A. 1981. Thrips. Pages 176–186 in: Grape Pest Management. D. L. Flaherty, F. L. Jensen, A. N. Kasimatis, H. Kido, and W. J. Moller, eds. Publ. 4105. Division of Agricultural Sciences, University of California, Berkeley. 312 pp.

Yokoyama, V. Y. 1977. *Frankliniella occidentalis* and scars on table grapes. Environ. Entomol. 6:25–29.

(Prepared by D. L. Flaherty and L. T. Wilson)

Leafhoppers and Treehoppers

Leafhoppers and treehoppers are especially important insects to consider in a discussion of grape diseases because injuries they cause by feeding and oviposition on vines are often confused with symptoms produced by diseases. In addition, some species in this insect group serve as vectors of the Pierce's disease bacterium and the *flavescence dorée* causal agent.

Few insects have plagued grape growers during the last century more than leafhoppers. Leafhoppers (Cicadellidae, subfamily Cicadellinae) that attack grapes in North America include a complex of *Erythroneura* species and the potato leafhopper, *Empoasca fabae* (Harris). Important leafhoppers that attack grapes in Eurasia include *Empoasca vitis* (Gothe), *Jacobiasca libyca* (Bergevin & Zanon), *Zygina rhamni* Ferrari, *Arboridia dalmatina* (Novak & Wagner), and *Scaphoideus titanus* Ball (syn. *S. littoralis* Ball) (vector of the *flavescence dorée* pathogen).

Erythroneura Leafhoppers

Important *Erythroneura* species included in the grape leafhopper complex are *E. calycula* McAtee, *E. comes* (Say) (eastern grape leafhopper), *E. coloradensis* Gillette, *E. elegantula* Osborn (western grape leafhopper), *E. maculator* Gillette, *E. tricincta* Fitch (threebanded leafhopper), *E. variabilis* Beamer (variegated leafhopper), *E. vitifex* Fitch, *E. vitis* (Harris), *E. vulnerata* Fitch, and *E. ziczac* Walsh (Virginiacreeper leafhopper).

Vine injury from *Erythroneura* leafhoppers (and that from *Z. rhamni* and *A. dalmatina*) appears first as a white speckling on the upper leaf surface (Plate 123) resembling virus infection. This speckling is caused by nymphs and adults piercing the mesophyll tissue from the underside of the leaf and emptying the contents of groups of palisade cells. The speckling is at first limited to the areas along the main veins and later spreads over the leaf blade and turns a blotchy pale yellow or whitish yellow. The greatest injury occurs on the basal one-third of leaves on a shoot. Leaves may become necrotic and drop. In severe infestations, partial defoliation may occur, which leads to reduced fruit quality and vine vigor. However, research in California and Pennsylvania has indicated that mature, vigorous vines can tolerate high densities of leafhoppers.

The liquid excrement of *Erythroneura* leafhoppers settles onto leaves and berries and may trap dust particles or support the growth of a sooty mold.

Empoasca Leafhoppers

The potato leafhopper, *Empoasca fabae*, breeds during the winter months in the Gulf Coast states. Adults increase in numbers during March and April and then migrate northward, becoming distributed over most of the United States. The leafhopper does not migrate south in autumn. Those in the colder regions perish with the occurrence of the first hard frost.

Empoasca vitis is widely distributed in Europe. In Italy it produces three generations per year and overwinters as adults on conifers.

Grape leaves injured by the potato leafhopper are mottled, with yellowed margins that roll or curl downward (Plate 124) similar to leafroll and grapevine yellows. Chlorotic areas may become necrotic late in the growing season.

Phloem feeders such as *Empoasca vitis* cause leaf vein browning, downward rolling, thickening, bright coloration, angular reddening (Plate 125) or yellowing, and occasionally a marginal leaf burn that might be confused with leafroll or grapevine yellows. Severe foliar injury inhibits the normal hardening of shoots into canes. Injury by *J. libyca* is similar and occurs in the hotter areas of Eurasia.

Empoasca spp. can injure grapes by mechanically injuring and blocking vascular tissue, by injecting toxic enzymatic secretions, or by a combination of both mechanisms.

Sharpshooter Leafhoppers

Sharpshooters (Cicadellidae, subfamily Tettigellinae) differ physically from the leafhoppers discussed above in that the simple eyes are positioned differently on the head. Their chief importance in vineyards is the transmission of Pierce's disease. Some 20 species of sharpshooters are vectors of the Pierce's disease bacterium throughout the range of this disease. Feeding by sharpshooters on leaves and cluster stems results in wilting and eventual necrosis of the affected parts.

Treehoppers

Common treehoppers (Membracidae) that feed on grapes include *Stictocephala bisonia* Kopp & Yonke (buffalo treehopper), *Enchenopa binotata* (Say) (twomarked treehopper), and *Spissistilus festinus* (Say) (threecornered alfalfa hopper). The treehoppers injure grapevines by feeding directly on the shoot and by depositing eggs therein. Shoots fed upon by the buffalo treehopper show a brownish girdling involving the epidermis and cortex and a thickening, reddening or yellowing, and downward rolling of the leaves distal to the girdle (Plate 126). These symptoms resemble leafroll and grapevine yellows. Some treehoppers create a ring of feeding punctures around the shoot, resulting in an enlarged callus just above the wound.

Selected References

Beirne, B. P. 1956. Leafhoppers (Homoptera: Cicadellidae) of Canada and Alaska. Can. Entomol. Vol. 88, Suppl. 2. 150 pp.

Bournier, A. 1976. Grape insects. Annu. Rev. Entomol. 22:355–376.

DeLong, D. M. 1948. The leafhoppers, or Cicadellidae of Illinois (Eurymelinae - Balcluthinae). Ill. Nat. Hist. Surv. Bull. 24:91–376.

Jubb, G. L., Jr., Danko, L., and Haeseler, C. W. 1983. Impact of *Erythroneura comes* Say (Homoptera: Cicadellidae) on caged 'Concord' grapevines. Environ. Entomol. 12:1576–1580.

McGiffen, K. C., and Neunzig, H. H. 1985. A guide to the identification and biology of insects feeding on muscadine and bunch grapes in

North Carolina. N.C. Agric. Res. Serv. Bull. 470. 93 pp.
Smith, F. F., and Poos, F. W. 1931. The feeding habits of some leafhoppers of the genus *Empoasca*. J. Agric. Res. 43:267–285.
Vidano, C., and Arzone, A. 1983. Biotaxonomy and epidemiology of Typhlocybinae on vine. Pages 75–85 in: Proceedings of the 1st International Workshop on Biotaxonomy, Classification and Biology of Leafhoppers and Planthoppers (Auchenorrhyncha) of Economic Importance. W. J. Knight et al, eds. Commonwealth Institute of Entomology, London. 500 pp.

(Prepared by G. L. Jubb, Jr.)

Phylloxera

The grape phylloxera, *Daktulosphaira vitifoliae* (Fitch) (Homoptera:Phylloxeridae), is an aphidlike insect of worldwide distribution. Its only known host is grape. Phylloxera is thought to have originated in eastern North America and moved from there to Europe and other grape-growing regions in the last half of the 19th century.

Individual phylloxera are minute and difficult to see with the unaided eye. However, because they are found in groups, both on roots and in foliar galls, their yellow or yellow-green color makes them conspicuous. Definitive identifications can be made in the field with a ×10 hand lens.

The life cycle of phylloxera is complex; some individuals feed on roots and others feed on foliage. The root form may be found in soil at any depth where grape roots are found. Soil that has a high clay content and cracks when it dries is a better medium for phylloxera than soil with high sand content which does not crack. Soil temperatures above 32° C are lethal to phylloxera eggs, as is standing water.

The foliar form of phylloxera is frequently found on the foliage of susceptible vines, both wild and cultivated, in humid viticultural regions. Leaf galls are rarely observed in arid areas.

Root Form

Symptoms of the root form of phylloxera appear as population densities become high and water stress is encountered in midsummer. Shoot growth and grape production are reduced. Under water stress and several years of high phylloxera infestations, plants die. Phylloxera damage appears initially as a few dead or declining contiguous vines in a vineyard. During subsequent years the area containing dead and declining vines enlarges somewhat concentrically.

Phylloxera feed by inserting their mouthparts into root or leaf tissue. Root or leaf galls form around the insertion point. Formation of the gall is vital to the nutrition of the growing phylloxera. The galling on roots is very different from that on leaves. If the feeding site is a mature, hardened root, the gall is called a tuberosity; if the feeding site is a rootlet, the gall is a bent, club-shaped area called a nodosity (Plate 127), which resembles galls caused by nematodes. Nodosities are easy to spot in soil around infested grape plants. They tend to be highly sensitive to rot organisms and do not last long enough to facilitate phylloxera development. Growth of the phylloxera population therefore depends on the formation of tuberosities. Dense populations of phylloxera give mature roots a blackened, rotted appearance.

Once the plant begins to decline, phylloxera populations decrease precipitously. Phylloxera are difficult to find on vines that have been severely stunted as a result of previous populations.

Foliar Form

Foliar phylloxera cause conspicuous spherical galls that protrude on the lower surface of leaves (Plate 128). When abundant, they may even gall young shoots and tendrils. The impact of foliar infestations is threefold: in the autumn the foliage turns color and falls off prematurely; plant vigor is reduced because of loss of photosynthetic capacity; and infestations of the leaf form lead to dispersal of the pest. If left unabated, some crawlers move from leaf galls to the roots, establishing new infestations or increasing existing infestations.

Selected References

Davidson, W. M., and Hougaret, R. L. 1921. The Grape Phylloxera in California. U.S. Dep. Agric. Bull. 903. 128 pp.
Granett, J., Bisabri-Ershadi, B., and Carey, J. 1983. Life tables of phylloxera on resistant and susceptible grape rootstocks. Entomol. Exp. Appl. 34:13–19.
Ordish, G. 1979. The Great Wine Blight. J. M. Dent and Sons, London. 233 pp.
Riley, C. V. 1874. The grape phylloxera (*Phylloxera vastatrix* Planchon). Pages 30–87 in: 6th Annual Report of Noxious, Beneficial and Other Insects of the State of Missouri. Jefferson City, MO. 169 pp.
Williams, R. N. 1979. Foliar and subsurface insecticidal applications to control aerial form of the grape phylloxera. J. Econ. Entomol. 72:407–410.

(Prepared by R. N. Williams and J. Granett)

Part III. Disorders Caused by Abiotic Factors

Chimeras

Chimeras or "bud sports" may suddenly appear within populations of perennial plant cultivars. They are new genotypes resulting from point mutations in meristems. The growth habit and characteristics of a chimera often resemble sudden infection by a pathogen.

Chimeras are usually stable and can be propagated asexually just like any other perennial plant genotype. Chimeras are occasionally selected as new cultivars with a new description. Bonnet de Retord and Sultanina Marble are two variegations that are mutant forms of older cultivars. Although they can be propagated by buds or cuttings, chimeras do not produce infections when they are grafted onto healthy plants. The fact that no transmissible agent is present in chimeric vines distinguishes them from virus-infected vines, which they may resemble.

Chimeras are found everywhere grapevines are grown commercially. At least four chimeras occur in grapes. Variegation is one common chimera in grapevines. With this chimera, a mosaic pattern spreads into new tissues and organs that develop from the point of mutation. The pattern may be separated from normal tissue by a distinct line or break in the color pattern within a single leaf (Plate 129). A spot within a leaf, a fruit cluster (Plate 130), a single shoot, a major part of the vine, or the entire vine may show the mosaic pattern, depending on the position of the mutation within the bud or meristem and the stage of development of the vine when mutation takes place. If a variegated shoot or cane is asexually propagated, the chimera persists in the daughter vines.

Fasciation is another common chimera in grapevines. The shoots, petioles, or cluster pedicels develop with a flattened shape (Plate 131) rather than the normal roundish shape. This can appear on any part of the vine. The mutation seems to occur with greater frequency in some cultivars than in others and is especially common in Petite Sirah (Durif). Genetic fasciations have frequently been confused with fasciations induced by grapevine fanleaf virus infection.

Witches'-broom is less common than either variegation or fasciation. In this chimera all buds developing from the original mutation seem to grow a short time after they are initiated and develop into a bushy growth covering part of the vine. Other buds on the same vine produce normal growth. Individual shoots that do develop in the broomed tissues do not mature, and no fruit clusters develop on these shoots. Leaves are smaller and remain green longer in the fall than normal leaves. Cuttings made from either green or dormant wood from affected canes will not root, so this chimera is normally not propagated asexually. Buds from affected canes can be grafted experimentally to a normal rootstock, where they continue to develop as a witches'-broom. The rootstock does not become affected.

The fourth chimera has not been given a formal name, probably because it is only rarely found in commercial vineyards. It is manifested chiefly as distorted growth of leaves (Plate 132). Apparently only one of the two outer layers in a bud meristem mutates, which results in a mix of normal and affected tissue within the shoot that subsequently develops. The shoot appears to grow normally, but leaves are grossly misshapen and are smaller than normal leaves. Sinuses of the leaf blade are reduced, and the color of the blade indicates that some chlorophyll-bearing tissues do not develop. The shoot is not greatly reduced in size. Cuttings made from affected canes root and produce vines with distorted leaves but no commercial fruit production. Buds from affected canes that are grafted onto healthy plants grow out as affected shoots but do not transmit any agent to the healthy plant. Normal-appearing explant lines have been experimentally developed from this chimera following prolonged heat treatments.

Chimeras observed on young vines in a new vineyard should be removed by pruning back to normal tissue. This practice effectively eliminates most chimeras in grapevines. It does not eliminate fasciations, however, because these seem to result from mutable genes, and the rate of mutation cannot be controlled.

Selected References

Dermen, H. 1947. Histogenesis of bud sports and variegations. Proc. Am. Soc. Hortic. Sci. 50:51–73.

Rives, M. 1970. Chimaeras and the like. Pages 255–256 in: Virus Diseases of Small Fruits and Grapevines (a Handbook). N. W. Frazier, ed. Division of Agricultural Sciences, University of California, Berkeley. 290 pp.

(Prepared by A. C. Goheen)

Nutritional Disorders

Nutritional disorders in grapevines are manifested by changes in the shape, color, chemical composition, performance, and attainable age of individual organs or the whole vine. The visible symptoms provide clues about their cause—the deficiency or the excess of one or more nutrients—but the appearance of the whole vine and of the vineyard may also aid in the diagnosis. Soil and petiole analysis can confirm nutrient imbalance. These data, combined with a knowledge of the soil, the sensitivity of the grape cultivar, and the environmental conditions, increase the accuracy and significance of the diagnosis.

Nitrogen

When nitrogen becomes limiting, vine foliage turns pale green and then yellow. Young shoots, petioles, and cluster stems become pink or red. Shoot growth is greatly reduced.

Sometimes light brown islands of dead tissue appear between the main veins of the basal leaves, and in extreme cases the leaf blades may wilt and abscise. Berries may be small. No specific patterns of organ malformation are associated with nitrogen deficiency. In commercial plantings the first symptoms may appear after berry *véraison* (color change) because nitrogen is translocated from the leaves near the clusters into the berries.

During long periods of cool, wet weather, vines may develop "cool-weather chlorosis," which can be confused with nitrogen deficiency. Low temperatures reduce chlorophyll synthesis, and the chlorosis disappears as soon as temperatures rise. Mechanical injuries and root damage from soil pests such as nematodes and phylloxera may hinder the absorption and transport of nutrients and lead to symptoms that may be confused with nitrogen deficiency.

Excess nitrogen increases growth; internodes become long and large, and leaf blades become deep green, thick, and sometimes cupped. Shoot growth may be excessive and prolonged. Grapevines are more tolerant of high nitrogen when other micronutrients and macronutrients are in adequate supply and when the trellis system allows for adequate leaf exposure to light.

Phosphorus

Phosphorus deficiency is an uncommon nutritional problem in most grape production areas. Thus most reports are based on symptoms obtained in nutrient culture. In such cases shoot and root growth are reduced and leaves are small and dark green. Leaf margins turn down without rolling. Under severe deficiency, leaves develop red, punctiform spots. Only a few experimental results report crop increases achieved through phosphorus fertilization. A history of excess phosphorus fertilization is more common. Excess phosphorus may induce zinc and iron deficiencies. Phosphorus deficiency may be one component of the nutrient imbalance brought about by low soil pH called *Säureschäden* (discussed below).

Potassium

The symptoms of potassium deficiency vary with the stage of development of the leaf blades when the potassium content in the tissues falls below the critical level. During the early part of the growing season, the leaves lighten in color in areas, and a few necrotic spots appear along the margin of young blades (Plate 133). During dry weather, necrotic areas varying in form, number, and size develop sporadically in the interveinal tissue. Leaf margins dry and roll up or down, and blades become distorted and ruffled (Plate 134). In late summer, the surface of older leaves at the base of shoots, receiving direct sunlight, becomes violet brown to dark brown ("black leaf"), especially near the clusters. Black leaf begins interveinally but can progress to completely cover the upper leaf surface (Plate 135). The leaf browning is especially pronounced on heavily cropped vines because after *véraison,* maturing berries become a potassium "sink." Potassium deficiency symptoms are more common in dry years.

Magnesium

Magnesium deficiency occurs on light, acid soils with a low magnesium content, sandy soils with a relatively high potassium content, and calcareous (high calcium carbonate) soils. High applications of potassium or ammonium can induce magnesium deficiency symptoms on soils that originally contained enough magnesium.

Magnesium deficiency takes two forms: early in the season, leaf necrosis dominates; during the summer and fall, interveinal yellowing is the major symptom. The first changes usually appear before bloom and may be small, brown-green spots near the margins and in the interveinal tissues of young, growing leaves (Plate 136). The elliptic or oval necroses form chains, which run within a few millimeters parallel to the leaf margin. In early to late summer, a brightening of the tissue appears between the main veins and, with increasing yellowing, proceeds from the leaf margin toward the attachment of the petiole in the form of a wedge between the primary and secondary veins (Plate 137). Magnesium deficiency at this time can be differentiated from deficiencies of other nutrients such as manganese, potassium, zinc, and boron by its striking straw yellow chlorosis pattern and its appearance first on basal leaves.

Calcium

Calcium deficiency appears occasionally on quartz gravel soils that are strongly acid ($pH < 4.5$). At first a narrow necrotic border appears at the margin of leaves and moves in steps toward the point of attachment of the petiole. Dark brown pimples up to 1 mm in diameter may develop on the primary bark of the internodes. The growing clusters dry up starting from the tip, which resembles severe stem necrosis.

Stem Necrosis (*Stiellähme*)

Shortly after *véraison* shallow or somewhat hollow (concave), usually elongated, brown to black necrotic areas appear on the rachis and its branches on vines with stem necrosis, a physiological condition frequently but not invariably associated with magnesium or calcium deficiency (Plate 138). Less frequently, circular, dark brown constrictions develop on the pedicels. In Europe, calcium chloride, magnesium chloride, and magnesium sulfate sprays are prescribed as a preventive treatment. However, a similar disorder in California ("water berry") and Chile (*palo negro*) is associated with higher nitrogen and ammonium in the affected tissues but has no relationship to low calcium or magnesium.

Säureschäden

The name *Säureschäden* is used to describe leaf symptoms on vines growing in extremely acid (pH 3.4–4.5) soils low in calcium and magnesium (Plate 139). The symptoms are associated with deficiencies of calcium, magnesium, and phosphorus. Excess potassium, aluminum, and especially manganese are also involved.

Shortly after bloom the margins of older leaves turn yellow or light brown. Brown spots develop along the margin and coalesce into larger, elongated, rust brown areas with irregular outlines. Red cultivars may have bright red spots. As in calcium deficiency, the damaged area dies slowly. In dry weather *Säureschäden* progresses rapidly. Frequently, the symptoms appear together with those of magnesium deficiency. Basal leaves may abscise. In early summer on acid soils, a bewildering variety of combinations and transitions of these symptoms can be observed. The clusters on grapevines with *Säureschäden* rarely mature fully, and the poorly matured canes do not tolerate low winter temperatures.

Iron

Iron deficiency (also called iron chlorosis, lime chlorosis, and lime-induced chlorosis) is especially common in regions with high-lime soils. The loss of chlorophyll starts between the small leaf veins. Fading begins at the leaf margins but progresses interveinally (Plate 140). Leaves may dry and fall. Set may be reduced. In cold, wet soils, transient iron deficiency symptoms may develop in the spring.

Manganese

Manganese deficiency is observed mainly on alkaline, sandy soils high in humus and on manganese-poor lime soils. In early summer the leaves at the base of shoots start to pale, and shortly afterward small, yellow spots appear in the interveinal tissue. The spots show a mosaiclike arrangement and are bordered by the smallest green veins. Only a very small seam along first- and second-order veins remains green (Plate 141). The symptoms are more severe on leaves exposed to the sun than on shaded leaves. Leaves are not malformed as with zinc deficiency. Advanced manganese deficiency affects the growth of shoots, leaves, and berries, and the maturation of the clusters is delayed. The symptoms of manganese deficiency on soils that contain lime are frequently concealed by severe yellowing from lime chlorosis appearing simultaneously.

In highly acid soils or soils rich in the element, manganese can be in excess (see *Säureschäden).*

Zinc

Zinc may be deficient in soils with low zinc levels such as very sandy soils or where topsoil has been removed. Zinc availability can also be affected by high phosphorus, which precipitates zinc as insoluble zinc phosphate, or in high-pH soils.

The first symptoms of zinc deficiency are small leaf blades with opened petiolar sinuses and sharp teeth (Plate 142). The leaf blades are commonly asymmetric, one half of the leaf being larger than the other. The interveinal areas turn light green to yellow in a mosaic pattern and can become reddish in black and red cultivars. The leaf veins also become clear, with narrow borders of green. The more advanced chlorotic areas can become necrotic. The intensity and distinctness of the symptoms vary somewhat among cultivars. Zinc deficiency reduces yields because vines produce fewer seeds and smaller berries (Plate 143). Zinc deficiency symptoms may resemble fanleaf degeneration.

Boron

Boron deficiency can have dramatic effects on vine growth and fruiting. It is observed especially on strongly acid (pH 3.5–4.5) soils and less frequently on neutral and alkaline soils (pH 7–8.5). Drought in the root area effectively hinders the absorption of boron. Vineyards in areas with very high rainfall and those irrigated with water devoid of boron are also susceptible, especially in easily leached, sandy soils.

The first symptoms appear on tendrils near the shoot tip before bloom. Dark, knotty bulges form and become necrotic. The distal portions dry up, and flower clusters die. During the period of rapid shoot growth, younger internodes swell slightly at one or several places and the pith becomes necrotic (Plate 144). Usually the part of the shoot distal to the swelling dies. Leaves have short, thick petioles sometimes showing longitudinal lesions or necrotic caverns. The blades have odd shapes and show interveinal chlorosis or necrosis (Plate 145). In the second season, buds formed on boron-deficient vines may produce short, bushy, branched, sterile shoots. Boron deficiency also affects the development of berries and clusters. Only a few seeded berries set, and small, seedless berries develop ("pumpkin and peas" symptom) (Plate 146). Roots remain short, are thickened, and swell into knots that break open longitudinally.

An oversupply of boron affects the development of all aboveground parts of the vine. The younger leaves can be severely malformed. Necrosis develops on the tips of the serrations of older leaves and progresses from the margin to interveinal areas (Plate 147). The tip growth of the main shoots decreases in favor of the lateral shoots, which produces vines that look weak and bushy.

Selected References

Bovey, R., Gärtel, W., Hewitt, W. B., Martelli, G. P., and Vuittenez, A. 1980. Virus and Virus-Like Diseases of Grapevines. Editions Payot, Lausanne. 181 pp.

Champagnol, F. 1984. Éléments de Physiologie de la Vigne et de Viticulture Générale. Editions Champagnol, Saint-Gely-du-Fesc, France. 351 pp.

Christensen, L. P., Kasimatis, A. N., and Jensen, F. L. 1978. Grapevine nutrition and fertilization in the San Joaquin Valley. Div. Agric. Sci. Univ. Calif. Publ. 4087. 40 pp.

Cook, J. A. 1966. Grape nutrition. Pages 777–812 in: Nutrition of Fruit Crops. N. F. Childers, ed. Horticultural Publications, Rutgers University, New Brunswick, NJ. 888 pp.

Fregoni, M. 1980. Nutrizione e Fertilizzazione della Vite. Edagricole, Bologna. 418 pp.

Gärtel, W. 1974. Die Mikronährstoffe - ihre Bedeutung für die Rebenernahrung unter besonderer Berücksichtigung der Mangel- und überschusserscheinungen. Weinberg Keller 21:435–508.

Smith, C. R., Shaulis, N., and Cook, J. A. 1964. Nutrient deficiencies in small fruits and grapes. Pages 327–357 in: Hunger Signs in Crops. H. B. Sprague, ed. David McKay Co., New York. 461 pp.

(Prepared by W. Gärtel; translated by H. O. Amberg)

Environmental Stress

Drought

The first response of grapevines subjected to water stress is the cessation of shoot elongation. Stressed plants are small, do not fill their trellis with leaves, and tend to have loose clusters with small berries. Fruit set may be reduced on vines stressed before or soon after bloom. Under prolonged and severe drought conditions, basal leaves first show marginal necrosis and then senesce and abscise (Plate 148). Periderm tends to form very early on the shoots of stressed vines. In nonirrigated areas the whole vineyard often shows signs of stress. However, and especially in irrigated vineyards, stress may be observed only on vines growing in isolated areas where the soil has very low water-holding capacity, on eroded knolls where the soil is shallow, or in areas where root growth is restricted by phylloxera, nematodes, or other local stresses such as a persistently high water table, which restricts root depth.

Excess Water

Prolonged flooding after growth begins kills roots by depriving them of soil oxygen. Vine growth is restricted, and plants often show symptoms of drought. The problem is usually associated with low or poorly drained areas or with vines near broken irrigation lines and with dead, decaying roots. If water is in abundant supply, as when the water table is stable and near the surface, shoot growth may be excessive and prolonged. Shoots on the excessively vigorous vines may fail to mature and may be killed by the first low temperature in fall or early winter.

Heat

Soon after exposure to damaging high temperatures, nonlignified portions of shoots wilt and discolor (Plate 149). Subsequently, the pith may dry and separate, as on vines struck by lightning (see Plate 152). Shoots then turn brown and dry up. Succulent tissues such as shoot tips, young leaves, and tendrils are most susceptible to heat injury. Berries are also damaged by excessive temperature (Plate 150), most often only on the top and western side of clusters exposed to the sun. Just parts of berries may be damaged (Plate 151) (sometimes referred to as Almeria spot), or whole berries may shrivel and turn brown (Plate 150). Berries and leaves that develop in the shade tend to be more sensitive to heat injury than are organs that develop in the sun. If shaded tissues are suddenly exposed to direct sunlight by summer pruning, leaf removal, or shoot positioning, sunburn frequently results. A sudden change from cool to hot weather increases the potential of sunburn because of poor acclimation of the tissues to the higher temperatures.

Lightning

Lightning damage to vines can be difficult to diagnose because of the variable and general nature of the vine response. In some cases the only response is bronzing and necrosis of leaves, which is reminiscent of chemical burns. In other cases the entire vine or vine row can collapse and die. The pith of shoots may dry and separate (Plate 152). Lightning strikes are indicated by the pattern of injury in the vineyard. Single whole rows are often damaged when lightning hits trellis wires. In nontrellised vineyards single vines or a small area of vines is usually affected. Evidence of lightning on the trellis or vine stakes, such as discolored wire or shattered vine posts or vine stakes (Plate 153), sometimes helps to confirm the diagnosis more than observations of the vines themselves.

Winter Injury

Buds and phloem of dormant vines are the tissues most sensitive to cold damage. Frequently the damage is restricted to low-lying areas of the vineyard or to parts of the vineyard where vine maturity the previous season was restricted by excessively vigorous growth or by crop, pest, or air pollution stress.

Bud injury can be detected while vines are still dormant by sectioning buds that have been thawed for more than 24 hours (Plate 154). Injured shoot primordia appear dark brown to black rather than the normal pale green. Shoot growth on vines with bud injury is sparse and irregular (Plate 155). Many buds fail to grow, and secondary and tertiary buds tend to begin developing later than healthy primary buds. In some cases the shoot primordium survives but the basal leaf primordia are injured (Plate 156). These leaves are often very small, can be malformed, and are frequently rugose with irregular, patchy chlorosis. These symptoms can be confused with those of virus diseases or with herbicide injury.

Phloem injury is also visible soon after the tissue thaws (Plate 157). Because temperatures are lowest near the ground, phloem injury is often found only on the trunk near the ground or snow line. On such vines shoot growth may appear normal for much of the season. Often the only sign of injury occurs when the shoots collapse from heat stress that results when the injured trunk cannot supply adequate water to meet the evaporative demand (Plate 158). Vines with trunk injury often have a large number of suckers (shoots arising from near or below ground level), which grow faster than the normal shoots. Vines with winter-injured trunks may survive one year but grow so poorly that they fail to survive the subsequent winter. These vines often have large cracks in the trunk that form as the tissues dehydrate. Winter-injured trunks also tend to be sites for crown gall formation.

Spring Freeze

After buds begin to swell, vine tissues are much less tolerant of low temperatures. The younger, more succulent tissues are most susceptible to injury from spring freezes. Soon after thawing, frozen tissue collapses and turns brown (Plate 159). Cold injury in spring can cause malformation of tissues within the expanded bud at higher temperatures than would cause similar injury in midwinter (Plate 160). Such malformed leaves could be confused with symptoms of Phomopsis cane and leaf spot. Cold damage is often restricted to low-lying portions of the vineyard or to vine growth near the ground. Because vines damaged by spring freezes produce few shoots and little or no crop, regrowth may be very vigorous on such vines. This excessive regrowth may cause a witches'-broom pattern of shoot growth early in the season and a failure of vegetative tissues to fully mature late in the season.

Hail

Hail injures vines by contact. Early in the season, hail may break shoots, leaves, or clusters or damage portions of the internode (Plate 161). These "hail pecks" can heal over and look like small galls or insect feeding wounds. Hail later in the season can cause tattered leaves and split fruit. Berries injured during the early stages of their development either abscise or shrivel and turn brown. The shriveled berries can be confused with black rot mummies. Later the injured fruit may become infected with *Botrytis cinerea* and other pathogens. Fruit injured after *véraison* usually rot.

Wind and Sand

Vines growing in windy regions can exhibit stunted growth because of low photosynthetic rates during a major part of the day, which is assumed to be related to the vine's response to high evaporative demand created by the wind. Vines subjected to high winds may have broken shoots, which results in unusual branching patterns. When strong winds blow late in the season, fruit may be injured and may become prone to bunch rot.

Blowing sand can cause fewer shoots, stunted shoot growth, partial defoliation, and/or malformed leaves (Plate 162). Such damage is usually restricted to a few vines on the windward side of the vineyard.

Salt Toxicity

Vines growing in a saline environment may show injury from chloride toxicity. The primary symptom is marginal necrosis of mature leaves (Plate 163). The expression of salt toxicity can vary markedly in vineyards because of differential susceptibility of scions and differential ability of rootstocks to exclude the chloride ion. Vines can also be injured by windblown sea spray. In such cases the foliage of affected vines becomes necrotic. Some cultivars are tolerant of windblown salt; for instance, in one vineyard Cabernet Sauvignon vines were relatively unaffected while adjacent Chardonnay vines were essentially defoliated.

Sprinkle-irrigated vineyards are also occasionally injured by external salt. Irrigation with water containing more than 3 meq of sodium or chloride is especially hazardous. Sprinkling is more hazardous during periods of high evaporation, with slow sprinkler rotation, and when the new spring growth is young and tender. There is a range in varietal tolerance of irrigation water salt similar to that of windblown salt.

Selected References

Jordan, T. D., Pool, R. M., Zabadal, T. J., and Tomkins, J. P. 1980. Cultural practices for commercial vineyards. Cornell Univ. Misc. Bull. 111. 69 pp.

Pool, R. M., and Howard, G. E. 1984. Managing vineyards to survive low temperatures with some potential varieties for hardiness. Pages 184–197 in: The International Symposium on Cool Climate Viticulture and Enology. D. A. Heatherbell, P. B. Lombard, R. W. Bodyfelt, and S. F. Price, eds. Oregon State University, Corvallis. 540 pp.

Shaulis, N. J., Einset, J., and Pack, A. B. 1968. Growing cold-tender grape varieties. N.Y. State Agric. Exp. Stn. Bull. 821. 16 pp.

(Prepared by R. M. Pool)

Air Pollution

Air pollution damages grapevines in many grape-growing areas of the world, including Japan, China, Israel, Italy, France, Rumania, Switzerland, Germany, Australia, Canada, and the United States. Air pollutants result from industrial gases and particles, automobile exhaust gases, and agricultural chemicals (especially herbicides). For many years the major causes of air pollution injury to grapevines were hydrogen fluoride (or silicon tetrafluoride) and, to a much lesser extent, sulfur dioxide. Within the past 40 years, however, injury from photochemical oxidants, principally ozone, has been significant in grape production in the United States and Canada and is recognized to occur in other areas of the world. Other substances, such as heavy metals (especially cadmium, lead, copper, and zinc from metal smelting, incineration, and municipal wastes), chlorine and chloride (products of industry and incineration of chloride-containing plastics such as polyvinyl chloride), and herbicides, are potential environmental phytotoxins but have rarely been so severe as to limit grape production, except under extremely localized conditions (see also Pesticide Toxicity section).

Injury to plants from air pollutants is determined by the pollutant, the plant, exposure conditions, and the environment. For example, pollutant uptake by leaves or (occasionally) roots of plants depends on the form of the pollutant—whether a gas or a particle, whether soluble or insoluble—and its composition as well as on the concentration of the pollutant and the duration and frequency of exposure. Light intensity, relative humidity, air and soil temperature, soil moisture, nutritional status, and precipitation at the time of exposure are also important in determining the degree of injury a plant sustains from air pollutants. Finally, biological, genetic, and cultural factors, such as plant species or cultivar, stage of plant development, plant health or vigor, age of the plant, and viticultural practices, also affect the accumulation of a pollutant, its distribution in the plant, and its elimination by leaching, weathering, volatilization, or metabolism. Many of these environmental and biological variables can be as significant in determining effects on the plant as are the pollutants themselves.

Ozone

The most commonly observed photochemical oxidant injury in the field is caused by ozone. Oxidized organics, such as peroxyacetyl nitrate (PAN), cause silvering, bronzing, and necrosis, but on the lower rather than the upper epidermal surface. Research has been limited to experiments with ozone or the ambient photochemical complex; although effects may occur, little or no evidence has yet been brought forth to show that other components of the "oxidant complex," such as PAN, aldehydes, and ethylene, are associated with grape injury, either alone or combined with ozone.

Experiments and field observations in California and New York State indicate that ozone is a major phytotoxicant that may limit grape production in some areas of the United States. Reduced yields of exposed Thompson Seedless grapes in California were primarily the result of reduced fruit set. Sugar content of individual berries was also slightly reduced, but no significant impact on either berry size or acid content was found. Thus berry and vine growth, yield, and quality can be reduced in the field by ambient ozone concentrations, especially under conditions prevailing in southern and central California.

Symptoms. The complex of ozone-induced symptoms is commonly referred to as oxidant stipple. Small, brown to black, discrete lesions are confined to palisade cells of the upper leaf surface (Plate 164) in areas of the leaf bounded by the smallest veins. Where lesions have coalesced, the upper epidermal cells over the damaged palisade cells are often collapsed. These symptoms may be distinguished from other grape disorders such as potassium deficiency (black leaf) by the stippled appearance of the lesions and the undamaged veins of affected leaves. In addition, oxidant stipple appears first and is most severe on basal leaves (i.e., one through six), and microscopic examination reveals the damaged palisade cells. The primary lesions vary from 0.1 to 0.5 mm in diameter; large lesions up to 2 mm in diameter may be produced by coalescence. The typical stippled appearance is the result of aggregates of the primary lesions. When oxidant injury is severe, leaves may become yellowed or bronzed, senesce prematurely, and abscise. Older leaves and the oldest parts of younger leaves are most vulnerable to oxidant stipple. In most cultivars, the necrotic stipple lesions remain small. They may become enlarged in Blue Elba and Grenache and are often accompanied by anthocyanosis in New York Muscat.

Cultivar Sensitivity. Some cultivars are extremely sensitive to ozone, whereas others exhibit remarkable tolerance. However, these observations have concerned foliar lesions, and the relationship between the degree of sensitivity of foliage and the effect on fruit yield has not been established.

Among the American cultivars, Ives has been shown to be most sensitive to ozone injury, Concord and Niagara are intermediate in tolerance, and Isabella, Delaware, and Dutchess are the most tolerant. The interspecific hybrid cultivars are generally more tolerant of ozone than the American cultivars. Rosette and Vignoles are relatively sensitive, Marechal Foch and De Chaunac are intermediate in tolerance, and Ravat blanc, Seyval, and Villard blanc are classified as tolerant.

Oxidant stipple had become a common feature in the grape-growing areas of California as early as 1958. Carignane, Grenache, Palomino, and Pedro Ximenes appear to be more sensitive than Burger, Thompson Seedless (Sultanina), and Zinfandel. Symptoms also occur in *V. californica* and *V. girdiana.* In experimental chambers, 89% of the total leaf surface of Carignane was affected after exposure to unfiltered ambient air containing ozone, compared to 62% in Palomino, 43% in Blue Elba, and 9% in Thompson Seedless.

Amelioration. Sprays of the fungicides benomyl and triadimefon have been reported to reduce ozone injury. Maintenance of a relatively high nitrogen level, cultivation in soils with good drainage, and the use of cover crops are also effective in minimizing ozone injury to grape foliage.

Hydrogen Fluoride

Industrial emissions of compounds containing fluoride have been reported to injure grapevines in many important grape-growing regions. Reduced yields of many *V. vinifera* cultivars due to airborne gaseous fluoride have been reported from Germany, Italy, Switzerland, France, China, Australia, and the United States near certain industrial areas.

Grape foliage, especially of some *V. vinifera* cultivars, is very sensitive to airborne fluoride. The foliage can absorb and accumulate high concentrations of fluoride from the atmosphere, but translocation from leaf to fruit and direct uptake by the fruit from the atmosphere are not significant pathways. Grape foliage is not an exceptional accumulator of fluoride compared with many other species. The threshold concentration of accumulated fluoride for foliar injury is best determined under field conditions and appears to be 35–40 ppm in the most susceptible cultivars.

The effect on yield may depend less on the degree of foliar injury than on exposure conditions in one or more growing seasons and the stage of vine development. Studies in Australia and elsewhere have shown that limited foliar injury and/or limited accumulation of fluoride in the foliage (less than 35–40 ppm, depending on the cultivar) has not altered yield or quality of the fruit. Where foliar injury is severe or accumulation of fluoride is great, the degree of reduction of yield and quality may depend on the time of exposure in the growing season. But the cumulative effect over several years may reduce vine vigor and affect yield or quality.

Symptoms. The first symptom of fluoride injury is the appearance of a gray-green color at the margin of the expanding or recently expanded leaf. The affected area is supple but ultimately turns brown or reddish brown and is often separated from the green portion of the lamina by a darker brown, reddish brown, or purple band (Plate 165). A thin band of chlorotic tissue may occur as a transition zone between the dark band and the green area. Dark concentric bands also may appear in the necrotic areas from earlier exposures to fluoride.

Cultivar Sensitivity. Among commercial *V. vinifera* cultivars, Roter Gutedel, Mission, Mataro, and Burger have been classified as highly sensitive; Palomino, Zinfandel, and Alicante Bouschet are moderately sensitive; and Blue Elba, Carignane, and Grenache are moderately tolerant. Other species of *Vitis* are also susceptible to fluoride injury and are often used as biological indicators. The degree of sensitivity also varies with the season (or stage of development). In greenhouse fumigations, for example, Colombard is ranked as very sensitive and Pedro Ximenes as very tolerant at the time of the spring flush. But in early fall, Colombard becomes less sensitive and Pedro Ximenes more sensitive.

Amelioration. Application of lime or other calcium-containing compounds can protect plants from airborne fluoride injury. Bordeaux mixture, a combination of copper sulfate and lime, also affords nearly complete protection to leaves of Semillon.

Sulfur Dioxide

Exposure to sulfur dioxide can reduce vine and fruit growth, lower yield and quality, and cause premature defoliation, but at concentrations and durations that are generally higher than those normally found or expected in the field. It is clear from available information that grapes are generally more tolerant of sulfur dioxide than of ozone and fluoride. Evidence has been presented, however, that indicates that the presence of sulfur dioxide exacerbates oxidant stipple on leaves of many species. These air pollutants can occur in the atmosphere simultaneously or, more commonly, sequentially. Where combinations of the two pollutants have been tested at concentrations found in the ambient environment, no synergistic responses (defined as a greater-than-additive effect of the two pollutants together) on growth, fruit yield, or quality have been reported.

Sulfur dioxide can injure table grapes in storage (see Pesticide Toxicity section).

Symptoms. Young leaves of *V. labrusca* 'Fredonia' exposed to sulfur dioxide develop grayish brown lesions at the margin, tip, or interveinal areas (Plate 166). In fully expanded leaves, interveinal areas turn a dark grayish green then change to grayish brown. Veins remain green. Severely affected leaves often abscise. Midshoot leaves are more sensitive to injury than are terminal or mature, basal leaves.

Cultivar Sensitivity. In field experiments in California, Cabernet Sauvignon and White Riesling grapes exposed to relatively low concentrations of sulfur dioxide (0.06 ppm or less for more than 90% of the time) exhibited foliar symptoms. In New York State, exposure to about 0.10 ppm sulfur dioxide alone for 12 hr each day resulted in necrotic and chlorotic foliar symptoms in some years. Limited studies in Japan on acute sulfur dioxide toxicity indicated that Fredonia was the most sensitive cultivar, followed by Delaware (Kyoho), Neomuscat, Kyogei, and Koshu, in decreasing order. In Ontario, Canada, exposure of several cultivars to a relatively high concentration of sulfur dioxide (0.6 ppm for 6 hr/day for four days) resulted in essentially no injury to the interspecific hybrids Baco noir, Marechal Foch, De Chaunac, New York Muscat, Villard noir, and Le Commandant.

Selected References

Brewer, R. F., and Ashcroft, R. 1983. The effects of ambient oxidants on Thompson Seedless grapes. California Air Resources Board. 15 pp.

Brewer, R. F., McColloch, R. C., and Sutherland, F. H. 1957. Fluoride accumulation in foliage and fruit of wine grapes growing in the vicinity of heavy industry. Proc. Am. Soc. Hortic. Sci. 70:183–188.

Doley, D. 1984. Experimental analysis of fluoride susceptibility of grapevine (*Vitis vinifera* L.): Foliar fluoride accumulation in relation to ambient concentration and wind speed. New Phytol. 96:337–351.

Fujiwara, T. 1970. Sensitivity of grapevines to injury by atmospheric sulfur dioxide. J. Jpn. Soc. Hortic. Sci. 39:13–16. (English translation)

Heck, W. W., Taylor, O. C., Adams, R., Bingham, G., Miller, J., Preston, E., and Weinstein, L. 1982. Assessment of crop loss from ozone. J. Air Pollut. Control Assoc. 32:353–361.

Jacobson, J. S., and Hill, A. C., eds. 1970. Recognition of Air Pollution Injury to Vegetation: A Pictorial Atlas. Informative Report 1, TR-7, Agricultural Committee. Air Pollution Control Association, Pittsburgh, PA.

Kender W. J., and Musselman, R. C. 1976. Oxidant stipple: An air pollution problem of New York vineyards. N.Y. Food Life Sci. Bull. 9(4):6–8.

Middleton, J. T., Kendrick, J. B., and Darley, E. F. 1955. Airborne oxidants as plant-damaging agents. Proc. Natl. Air Pollut. Symp. 3:191–198.

Murray, F. 1983. Response of grapevines to fluoride under field conditions. J. Am. Soc. Hortic. Sci. 108:526–529.

Musselman, R. C., and Melious, R. E. 1984. Sensitivity of grape cultivars to ambient O_3. HortScience 19:657–659.

Musselman, R. C., and Taschenberg, E. F. 1985. Usefulness of vineyard fungicides as antioxidants for grapevines. Plant Dis. 69:406–408.

Musselman, R. C., Shaulis, N. J., and Kender, W. J. 1980. Damage to grapevines by fossil fuel wastes and pollutants. Search: Agric. No. 3. New York Agricultural Experiment Station, Geneva. 19 pp.

Musselman, R. C., Forsline, P. L., and Kender, W. J. 1985. Effect of sulfur dioxide and ambient O_3 on Concord grapevine growth and productivity. J. Am. Soc. Hortic. Sci. 110:882–888.

Richards, B. L., Middleton, J. T., and Hewitt, W. B. 1958. Air pollution with relation to agronomic crops: V. Oxidant stipple of grape. Agron. J. 50:559–561.

Thompson, C. R., Hensel, E., and Kats, G. 1969. Effects of photochemical air pollutants on Zinfandel grapes. HortScience 4:222–224.

Thompson, C. R., Kats, G., and Dawson, P. J. 1982. Low level effects of H_2S and SO_2 on grapevines, pear, and walnut trees. HortScience 17:233–235.

Weinstein, L. H. 1984. Effects of air pollution on grapevines. Vitis 23:274–303.

(Prepared by L. H. Weinstein)

Pesticide Toxicity

If pesticides (fungicides, insecticides, herbicides, or growth regulators) are applied improperly, they may injure grapevines and elicit a phytotoxic response. Phytotoxicity may result from an excessive application rate, the application of incompatible pesticide mixtures, application at a sensitive stage of vine development, application during or before unfavorable environmental conditions, or application to sensitive cultivars. Specific formulations of material used or the addition of adjuvants may trigger a phytotoxic response. Herbicide injury to grapevines may occur through absorption by roots or by leaves from either a direct application or from spray drift or volatiles. Preplant soil fumigants may injure a new vineyard if plants are set before residues of the fumigant have been adequately reduced.

Phytotoxicity symptoms vary depending on the chemical, its concentration, and the stage of vine development at the time of exposure (Table 3). Young, unfolding or expanding leaves are especially sensitive. Stunted shoots, malformed leaves, chlorotic leaves, or necrotic leaves may be observed. Berries sprayed during the early stages of their development are highly sensitive to injury, which may appear as russeting, scarring, or even cracking. Phytotoxicity symptoms caused by some pesticides may be easily confused with diseases—for example, paraquat injury and black rot, glyphosate injury and Eutypa dieback, and endosulfan injury and *Rotbrenner*—or with disorders with abiotic causal agents, such as magnesium deficiency. Injury may be associated with specific environmental parameters, such as high temperatures during or shortly after applications of sulfur or dinocap, or cool temperatures during applications of captan or vinclozolin.

Diagnosis of phytotoxicity is frequently possible only after a thorough analysis of the grower's spray record during the current season and in previous years, including accompanying weather records; an examination of neighboring crops and weeds within the vineyard; observations on the location of injured vines within the vineyard; and determination of the injured organs on individual vines. Injury symptoms in newly set vineyards may be related to previous crops and the pesticides used on those crops. Because of the complexity of diagnosing phytotoxic responses, it is frequently advisable to consult more than one specialist.

Selected References

Bolay, A., and Caccia, R. 1979. Effets des traitements cupriques sur le rougissement précoce du feuillage du cépage Merlot au Tessin. Rev. Suisse Vitic. Arboric. Hortic. 11:205–211.

Bovey, R., Gärtel, W., Hewitt, W. B., Martelli, G. P., and Vuittenez, A. 1980. Virus and Virus-Like Diseases of Grapevines. Editions Payot, Lausanne. 181 pp.

Clore, W. J., and Bruns, V. F. 1953. The sensitivity of the Concord grape to 2,4-D. Proc. Am. Soc. Hortic. Sci. 61:125–134.

Doster, M., and Sall, M. A. 1984. Phytotoxicity to grapevines of fenarimol and triadimefon. Am. J. Enol. Vitic. 35:97–99.

Haeseler, C. W., and Petersen, D. H. 1974. Effect of cupric hydroxide vineyard sprays on Concord grape yields and juice quality. Plant Dis. Rep. 58:486–489.

Kuck, K. H., and Scheinpflug, H. 1986. Biology of sterol-biosynthesis inhibiting fungicides. Pages 65–96 in: Chemistry of Plant Protection. Vol. 1. G. Haug and H. Hoffman, eds. Springer-Verlag, Berlin. 151 pp.

Pearson, R. C. 1986. Fungicides for disease control in grapes, advances in development. Pages 145–155 in: Fungicide Chemistry, Advances and Practical Applications. M. B. Green and D. A. Spilker, eds. American Chemical Society, Washington, DC. 173 pp.

Shaulis, N. J., Crowe, D. E., and Rogers, R. A. 1978. The relation of the phytotoxicity of glyphosate to its injury-free use in vineyards. Part 1. Glyphosate studies as a basis for injury-free use. Proc. Northeast. Weed Sci. Soc. 32:246–253.

Suit, R. F. 1948. Effect of copper injury on Concord grapes. Phytopathology 38:457–466.

Taschenberg, E. F., and Braun, A. J. 1960. Evaluation of phaltan for the control of grape diseases. Plant Dis. Rep. 44:560–565.

Taschenberg, E. F., and Shaulis, N. 1955. Effects of DDT-Bordeaux sprays and fertilizer programs on the growth and yield of Concord grapes. Proc. Am. Soc. Hortic. Sci. 66:201–208.

Tweedy, B. G. 1981. Inorganic sulfur as a fungicide. Residue Rev. 78:43–68.

Weaver, R. J. 1970. Some effects on grapevine of exogenous plant regulators and herbicides. Pages 247–254 in: Virus Diseases of Small Fruits and Grapevines. N. W. Frazier, ed. Division of Agricultural Sciences, University of California, Berkeley. 290 pp.

(Prepared by R. C. Pearson, R. M. Pool, and G. L. Jubb, Jr.)

TABLE 3. Symptoms of Pesticide Injury to Grapevines

Pesticide	Symptoms	Remarks
	Herbicides	
Glyphosate	First year: Leaves are arrow-shaped and very rugose, with or (more frequently) without interveinal chlorosis. Internodes may be shortened (Plate 167). Apical dominance may be broken, resulting in growth of numerous lateral shoots. Second year: Early shoot growth is stunted, with severely strapped leaves that can be confused with Eutypa dieback (see Plate 53).	Injury is usually due to translocation of glyphosphate applied to ground suckers or to drift from material applied under the trellis. Glyphosate moves with the photosynthates and may affect one or several shoots on a vine. Second-year symptoms usually develop from late-summer application the previous year.
Simazine	Basal leaves are usually most affected. Chlorotic patches originate at leaf margins. Leaves are not deformed (Plate 168).	Symptoms are caused by translocation from roots. Injury is usually associated with application of high rates to coarse soils low in organic matter and/or excessive application of water. Vines tolerate moderate levels of simazine injury.
Diuron	Primary symptom is yellow veins on otherwise normal leaves (Plate 169).	Most common symptoms are caused by translocation from roots. Injury is more common where root growth is shallow, soil is coarse, or excessive water is applied and is frequently observed in mulched vineyards. More unusual is contact injury, which can resemble simazine injury (Plate 168). Vines are tolerant of moderate diuron injury.
Phenoxy compounds (2,4-D, 2,4,5-T)	Leaves are strap-shaped, with greatly opened sinuses, highly elongated leaf enations, and pebbly rugosity (Plate 170). Foliar symptoms may be confused with fanleaf degeneration (see Plate 89). Depending on time of exposure, fruit set may be increased or unaffected. Application to developing berries may greatly delay or even prevent *véraison*.	Vines are very sensitive to long- and short-distance aerial drift. A few isolated affected leaves can be tolerated, but vines are damaged by very low levels. Volatile forms of 2,4-D should never be used in vineyard regions.
Dicamba	Affected leaves are cupped and have a distinct marginal band of restricted growth (Plate 171). Enations are very frilly. Translocation patterns are similar to those of 2,4-D.	Dicamba is rarely used in vineyards. Injury is usually associated with drift.
Aminotriazole	Mature leaves show marginal, patchy chlorosis and necrosis (Plate 172).	Injury is usually associated with drift from cornfields or with vineyards planted in previous cornfields.
Paraquat	Injury generally results from contact. Discrete spots initially are cleared, chlorotic areas and develop into necrotic spots (Plate 173). These spots can be confused with black rot (see Plate 21). Treated suckers show a general burn.	Injury is usually from vineyard application. Rarely, paraquat applied at ground level is absorbed and translocated to the canopy, where a general chlorosis can develop (Plate 174).
	Growth Regulators	
Gibberellic acid	First year: Clusters are elongated, with long pedicels, many shot berries, and retained ovaries. Rachis may be woody and twisted (Plate 175). Second year: Budbreak is reduced and irregular. Flower clusters are greatly reduced in number and size. Set is normal, but very few berries develop (Plate 176).	Injury usually occurs from misapplication to sensitive cultivars. Most seedless grapes tolerate gibberellic acid.
	Fungicides	
Sulfur	Injury on foliage generally appears as interveinal bleaching that becomes necrotic (Plate 177). Defoliation may occur. Injured fruit have brown to black scar tissue (Plate 178), which may eventually crack, depending on the stage of berry development at the time of exposure.	Injury is generally associated with exposure at temperatures over 30° C, but some cultivars, especially those of *V. labrusca* origin and some interspecific hybrids, are highly susceptible to sulfur injury even at lower temperatures.
Sulfur dioxide	Bleaching, the most common type of sulfur dioxide injury in stored table grapes, is most pronounced around the pedicel (Plate 179) or any break in the epidermis. Tissue under the bleached area dries out, collapses, and forms a depression. Changes in color are usually accompanied by objectionable changes in taste.	See Air Pollution section for effects of ambient sulfur dioxide on foliage.

TABLE 3. *(continued)*

Pesticide	Symptoms	Remarks
Copper	Injury ranges from slight bronzing of leaves to foliar necrosis (Plate 180) and defoliation. Fruit may exhibit black necrotic areas where tissue has been killed.	Injury may become more pronounced in the latter part of the season when heavy dew formation reactivates accumulated Cu^{2+} ions. So-called "fixed coppers" such as copper oxychloride are less phytotoxic than copper sulfate (the active ingredient in Bordeaux mixture), but all formulations require addition of lime as a partial safeguard. Even so, reduced vine growth and yield have been observed following extensive use of copper and lime.
Dinocap	Injury ranges from stunted, malformed leaves with necrotic sectors (Plate 181) to large, necrotic burn areas (Plate 182) on leaves exposed to the sun. Berry symptoms range from circular black spots to russeted areas where spray residues have dried.	Leaf malformation occurs when dinocap is applied to young, unfolding or expanding leaves of susceptible cultivars. Foliar burn is generally associated with exposure to temperatures over 30° C and is generally independent of cultivar. Emulsifiable concentrate formulations are more phytotoxic than wettable powder formulations.
Azoles (etaconazole, penconazole, triadimefon, etc.)	Internodes are shortened, and leaves are small, thick, dark green, and puckered (Plate 183). Leaf blades tend to curve downward.	Injury is usually associated with excessive rates of application.
Phenylamides (benalaxyl, metalaxyl, etc.)	Leaves develop marginal and interveinal chlorosis and eventual necrosis (Plate 184). Symptoms could be confused with simazine injury (Plate 168).	Injury is associated with application of excessive rates.
Dicarboximides (iprodione, vinclozolin, etc.)	Leaves are misshapen and puckered, with interveinal and marginal chlorosis (Plate 185).	Injury is associated with applications to expanding leaves during low temperatures.
Phthalimides (captan, folpet, etc.)	Leaves are misshapen, puckered, and rugose, and berries are scarred (Plate 186). Leaves may also show a general burning.	Injury is usually associated with spray applications during adverse weather. Phthalimide injures berries when sprays are applied under wet (foggy), cool (below 10° C) conditions. Foliar burning caused by folpet is associated with droughty soils and high temperatures.
Sodium arsenite	Buds can be injured from dormant applications, resulting in reduced and irregular budbreak and some crop loss. Upon dissection, dark brown, dead xylem tissue is observed beneath the leaf scars and lateral shoot scars at the base of nonemerging or slow-emerging buds (Plate 187). Injury occurs at the site of chemical absorption.	Late-dormancy treatments (two to three weeks before bud swell) are preferred to minimize injury. Sodium arsenite should not be applied within 10 days after pruning.
	Insecticides and Miticides	
Endosulfan	A rapid necrosis appears on portions of leaves that receive the heaviest spray deposit (Plate 188). Leaves may drop if injury is severe.	Injury is most severe on Baco noir, Cascade, Chancellor, Colobel, and Concord and is associated with hot weather at the time of application.
Phosalone	Leaves become chlorotic, develop a mosaic pattern, and crinkle; berries may become scarred.	Injury has been observed in California on Thompson Seedless vines under heat and moisture stress at application rates two to four times higher than recommended.
Propargite	Leaves take on a blotchy, purple-black color about one week after treatment; a necrosis develops in the blotchy areas three to four weeks after application.	Injury has been observed on Concord leaves when propargite was applied at two to four times the recommended rate. Injury may increase if weather is hot and dry during and immediately after application.

Part IV. Effects of Cultural Practices on Disease

Cultural practices can greatly influence the incidence and severity of disease in vineyards by modifying the vine microclimate so that it is more or less favorable to disease development, by altering the amount of disease inoculum present in the vineyard, by influencing the vines' resistance to disease, or by changing the vines' tolerance of the effects of disease.

Modifying the Vine Microclimate

Microclimate refers to the climate within the vine leaf canopy. In relation to disease, the important elements of the vine microclimate are relative humidity, ventilation, the temperature of the air and of the vine tissues, and the intensity and quality of light. In general, factors that increase relative humidity also increase fungal diseases. Factors that increase ventilation of the vine canopy generally reduce disease incidence and severity by lowering the humidity, shortening periods of wetness, and aiding spray penetration and coverage.

The temperature of vine tissues differs from that of the surrounding air. Tissue temperature is generally higher than air temperature during the day because of incident radiation received and lower than air temperature during the night because of heat loss by radiation. The expected increase in day temperature may either increase or reduce infection, depending on the temperature range that permits growth of the pathogen. Dew formation is related to the low night temperature of tissues free to radiate heat to the sky. Dew can be an important wetting mechanism for disease development.

Light intensity is important because it is responsible for the heating of vine tissues, but the effects of light quality are less well understood. Canopy interiors are characterized by low light and light that is enhanced in red as opposed to far-red radiation. Leaves that develop within the canopy are thinner, with a lower stomatal density and thinner cuticles, than leaves that develop on the surface of the canopy. Such leaves may be more susceptible to infection, and their function may be more impaired at a given level of infection, than leaves that receive direct sunlight. Light quality may also play a direct role in fungal spore germination.

Cultural practices can alter the microclimate directly or indirectly. An example of a direct effect is irrigation, which adds water to the vineyard ecosystem and raises the relative humidity, especially sprinkler or microjet irrigation that wets the foliage. Living mulches or cover crops may directly increase the relative humidity in the vineyard by evapotranspiration. Cover crops compete with vines for soil moisture and can dry out vineyard soils.

Site selection and use of windbreaks also directly affect microclimate. In the Northern Hemisphere, north-facing slopes receive less light than south-facing slopes and hence less radiation is available to alter the microclimate. Sloping sites encourage air movement and so increase ventilation and decrease radiant heating. Concave sites can import dense, wet air and lengthen wetting periods. Windbreaks can prevent shoot breakage, lower transpiration, and reduce the effects of wind on stomatal closure, which, if excessive, can reduce photosynthesis. However, they also decrease vineyard and vine ventilation and hence may prolong wetting periods.

Winter injury is known to increase vine susceptibility to crown gall bacteria. Thus, site selection and cultural practices that minimize such injury are important. Vines are sometimes covered or mounded with soil for protection in very cold areas. The insulating properties of soil reduce the extremes of tissue temperature and help prevent injury to protected aerial and root tissues.

Any cultural practice that alters vegetative growth and canopy density has an effect on vine microclimate. Most cultural practices are chosen primarily to enhance yield or fruit quality rather than to influence the microclimate. However, practices such as shoot thinning ("crown suckering"), leaf removal, berry and cluster thinning, pruning, and shoot positioning have a direct impact on vine microclimate. Increasing cluster thinning and decreasing pruning stimulate vegetative growth and hence reduce light exposure and ventilation within the canopy. Shoot thinning, leaf removal, and summer pruning are frequently done specifically to reduce canopy density, so as to increase fruit exposure to light, improve ventilation, and aid spray coverage. Shoot positioning is usually done to ensure canopy separation of divided canopies or to enhance light exposure of the renewal zone of the vine; it also decreases vegetative growth and canopy density and increases light exposure of fruit.

Altering Inoculum Levels

Growers can employ numerous cultural practices to reduce the inoculum level of pathogens in vineyards. These include destroying diseased material, removing alternative host reservoirs of inoculum, discouraging the presence of disease vectors in the vineyard, and preventing the introduction of diseased vines into the vineyard.

By removing infected tissues, growers prevent diseased material from releasing inoculum in the vineyard. Dormant-season pruning removes about 90% of the annual growth and also removes some older tissues of the vine. In the process, overwintering inoculum of some pathogens such as *Phomopsis viticola* is reduced. Piling prunings in areas adjacent to the vineyard is not advisable because inoculum is not removed from the ecosystem. Chopping or shredding prunings and incorporating them into the soil may help to reduce inoculum levels of aerial pathogens. Diseased trunks or arms should be removed from the vineyard and buried or burned. The value of annual pruning in disease control becomes apparent when minimal pruning is done. The retention of dead and diseased canes in the vine leads to increased infection. In New York State, minimally pruned vines have shown frequent fruit infection with *P. viticola,* a rare event in conventionally pruned vineyards.

Diseased grape berries (mummies), grape leaves, and plants other than grapes are important sources of inoculum that growers can destroy by cultural practices, for example by removing mummies from vines and the berm area and by cultivation. It is possible that the use of herbicides or nontillage instead of cultivation could lead to increased disease pressure.

Vineyard replantings require special consideration to minimize reinfection with existing pathogens. It is advisable to remove as many of the old roots as possible. A common practice is to follow vine and root cleanup with soil fumigation that kills roots as well as some pests. Vines infected with viruses present a special problem. At present, the only proven long-term solution is to plant a nonhost crop for at least 10 years before replanting with grapes. Recently released rootstocks resistant to dagger nematodes and immune to grapevine fanleaf virus offer promise for disease control in future replant vineyards. These approaches may also be beneficial in interrupting the carryover of other pathogens in replant sites.

Alternative hosts can be important sources of inoculum. In California, Pierce's disease is most prevalent in vineyards that border streams or other habitats that harbor infected weed host plants supporting vector insects and the causal bacterium. Removing these plants may help control the disease. Highly susceptible cultivars should not be planted near such habitats. In some locations, adjacent cultivated fields contain crops that are alternative hosts. In California's San Joaquin Valley, the greatest incidence of Pierce's disease usually occurs downwind from permanent pastures, weedy alfalfa fields, or other areas compatible to vectors, such as grasses along roadsides or in wet areas. Water grass, Bermuda grass, Italian rye, perennial rye, and fescue are especially attractive to green- and red-headed sharpshooter vectors.

Vineyard weeds can also serve as alternative hosts for some viruses. Plantains and dandelions in New York State are frequently infected with tomato and tobacco ringspot viruses. Vines near infected weeds can become infected when the nematode vector is present. In such cases, careful selection of herbicides or frequent cultivation may help prevent vine infection.

The importance of vine stock selection as a way to avoid disease cannot be overstated. Vineyards should be established only from stocks certified free of damaging viruses. Programs are now being developed to ensure that propagation stock is also free of other pathogens such as the crown gall bacterium.

Altering Disease Resistance of the Grapevine

The primary way to increase grapevine resistance to disease is through the selection of resistant (or tolerant) rootstocks or scions. However, other cultural practices can also contribute. Recent research in vine training has emphasized the establishment of light-porous leaf canopies. Leaves that develop in the light are thicker and have thicker protective cuticles than leaves that develop in poor light. Protective periderm develops earlier on shoots that develop in the light.

Excessive fertilization with nitrogen, abundant irrigation, or improper rootstock choice can lead to luxuriant growth of succulent, thin-leaved shoots that are prone to fungal infection. Proper water management can help prevent disease by reducing late-season shoot growth and inducing early periderm formation.

Careful choice of thinning and girdling practices and the use of growth regulators also can help increase resistance to disease. Excessively compact clusters cause berry splitting, which creates entry points for disease organisms, attracts insects, and encourages the spread of disease organisms among berries in the cluster. Recent research has shown that berry wax development is impeded at the points of berry contact in a cluster, and those contact points are more likely to become infected in vitro by applied *Botrytis* spores. Because grape berry size is strongly enhanced by soil moisture availability, deep, fine-textured soils that store considerable moisture are poor choices for cultivars with compact clusters of thin-skinned berries, such as Zinfandel and Chenin blanc. Cabernet Sauvignon is more suitable, if bunch rot is to be minimized. Careful insect and bird control programs also minimize feeding injuries that serve as entry points for disease organisms.

Use of pest-resistant plant material is an ideal way to lessen the susceptibility of a vineyard to disease. New nematode-resistant, virus-immune rootstocks offer the hope of preventing virus reinfection in replanted vineyards, and careful matching of the scion to the environment can do much to ward off disease. Disease-resistant cultivars are being developed in active breeding programs around the world, and advances in biotechnology offer hope for greater progress in this area in the future. In the meantime limited tolerance to disease is available within the cultivars of *V. vinifera.* Many disease problems can be minimized by careful selection of cultivar.

Increasing Vine Tolerance of Disease

A practical approach in vine disease management is to minimize economic injury rather than attempt to eradicate the disease. For example, slowly developing diseases of wood, such as Eutypa dieback and crown gall, can be managed by multiple-trunk systems. A "spare" trunk that is normally used to fruit half the vine space can be used to fruit the entire vine space should its companion trunk need to be removed because of infection. A replacement trunk can be established in one or two years by training a shoot from a sucker.

Another example of economic disease management is the use of gibberellic acid or other management techniques to decrease cluster compactness. Looser clusters reduce the spread but not the initiation of fungal infection within clusters. Application of gibberellic acid can reduce the number of berries per cluster and thus lessen cluster compactness. Decreasing pruning to increase node number, by increasing the number of clusters per vine and decreasing the number of berries per cluster and berry size, has been shown to reduce rot incidence in tight-clustered cultivars such as Chenin blanc.

Selected References

Boubals, D. 1982. Progress and problems in the control of fungus disease of grapevines in Europe. Pages 39–45 in: Grape and Wine Centennial Symposium Proceedings. D. A. Webb, ed. University of California, Davis. 398 pp.

Kliewer, W. M. 1982. Vineyard canopy management—A review. Pages 342–352 in: Grape and Wine Centennial Symposium Proceedings. D. A. Webb, ed. University of California, Davis. 398 pp.

Lynn, C. D., and Jensen, F. L. 1966. Thinning effects of bloomtime gibberellin sprays on Thompson Seedless table grapes. Am. J. Enol. Vitic. 17:283–289.

Savage, S. D., and Sall, M. A. 1984. Botrytis bunch rot of grapes: Influence of trellis type and canopy microclimate. Phytopathology 74:65–70.

Smart, R. E. 1985. Principles of grapevine canopy management manipulation with implications for yield and quality. A review. Am. J. Enol. Vitic. 36:230–239.

Wolf, T. K., Pool, R. M., and Mattick, L. R. 1986. Responses of young Chardonnay grapevines to shoot tipping, ethephon and basal leaf removal. Am. J. Enol. Vitic. 37:263–268.

(Prepared by R. M. Pool, A. N. Kasimatis, and L. P. Christensen)

Part V. Selection of Planting Material

Selection of Grapevines

Humans planted *V. vinifera* as a crop as far back as 5000 B.C. In the Near East and southern Europe, superior types were selected from wild vines and gradually domesticated. We know that grapes were vegetatively propagated in vineyards before the Christian Era. Most cultivars used in present-day viticulture were selected and named before any historical ampelographic documentation.

Problems developed in grape production in European vineyards after the introduction of grape diseases and pests from the New World between 1850 and 1878, especially the infestation of phylloxera in 1863. These diseases and pests were unknown in the old vineyards that had been planted before 1850. By 1878, only 20% of vines in Germany gave a satisfactory yield, 40% gave a very low yield, and 40% gave no yield at all. Poor performance led to early clonal selection, which predated the development of the modern disciplines of plant genetics, plant pathology, and viticulture and the wide use of phylloxera-resistant rootstocks.

Clonal selection during the late 19th century was chiefly either negative or positive mass selection until 1876, when individual mother vine selection was introduced in the Palatinate in Germany with the cultivar Sylvaner. In the original mass selection schemes, negative variants in the vineyard population were excluded from vegetative propagation and only vines with above-average performance were propagated. After the individual mother vine selection scheme was introduced, only one vine with top performance was selected for propagation. With this individual mother vine selection system for reconstructing vineyards, average must yields in the German wine industry have increased from 3,000 to 10,400 L/ha in the last 100 years without altering wine quality.

The individual mother vine selection scheme in Germany is a complex set of observations and replicated yield and quality tests requiring 20 years to complete. For example, 100 original clones (clones being defined as the family developing from a single mother vine selection) are selected initially as the 100 best vines from among 5,000 of a given cultivar growing on five different rootstocks (1,000 on each rootstock). During the course of testing, two clonal lines from the 100 selected clones of individual mother vines are finally registered with the German Federal Vine Cultivars Board for commercial propagation of vineyards.

During the tests, 30 characteristics are measured and used for selection. The tests measure stress resistance as well as quality factors in vines, must, and wine. Tests are optimized to select stable performance, that is, performance independent of site. Clones with small annual variation in yield are preferred. Clones that have stable quality with increasing yield are also preferred. Normally, wine quality is evaluated during the course of the tests.

Initially, every effort was made to pick clones free of dangerous viruses, but indexing did not start until the clonal selection process was well advanced. At that time the clonal selection scheme did not specifically call for virus disease-free rootstocks at any of the three phases of selection. In recent years all initial vines infected with grapevine fanleaf virus, leafroll, or fleck have been rejected before preselection. The selection scheme does not always call for tests to eliminate rupestris stem pitting or corky bark.

Experiments with White Riesling clones at the central station for clonal selection in Trier in the Federal Republic of Germany showed that heat treatments of propagation materials lead to improved performance and less variability in the clonal line. The results indicated that such treatments had beneficial effects and should be incorporated into the individual mother vine selection scheme in clonal propagation. Part but not all of the improvement could be attributed to elimination of leafroll and grapevine fanleaf virus. The experiments suggested that heat treatment might also eliminate additional virus infections.

These results demonstrate the need to use virus disease-free rootstocks and to conduct indexing tests for known virus diseases, including rupestris stem pitting and corky bark, at the earliest point in the individual mother vine clonal selection scheme. Rupestris stem pitting and corky bark are often latent in clonal lines of *V. vinifera*, and the effects, as measured in California, are slow decline in production and reduced quality.

The California selection scheme relies on early selection for commercial performance, trueness to type, and freedom from serious disease. Initial selections are made in a fashion similar to the German system, and indexing for serious virus disease problems is conducted before materials are registered with the Foundation Plant Materials Service in Davis. Indexing has now been incorporated into the German system. The major difference between the German and the Californian schemes is that performance characteristics of clonal lines in California are not measured over a 20-year period before the materials are registered. In either scheme the measurement of true somatic mutation effects will be possible only when virus disease infections are completely eliminated.

Selected References

Goheen, A. C. 1980. The California clean grape stock program. Calif. Agric. 34:15–16.

Schöffling, H. 1980. First results of a field trial on the performance of heat-treated and non-heat-treated White Riesling clones. Pages 311–320 in: Proceedings 7th Meeting ICVG, Niagara Falls, Canada. A. J. McGinnis, ed. Vineland Research Station, Vineland Station, Ontario. 355 pp.

Schöffling, H. 1984. Die Klonenselektion bei Ertragsrebesorten. Auswertungs- und Informationsdienst für Ernährung, Landwirtschaft und Forsten (AID), Bonn, Federal Republic of Germany. 24 pp.

(Prepared by H. Schöffling and A. C. Goheen)

Registration and Certification

Registering mother vines that are developed in research programs and increasing nursery stocks from such vines through a certification program have reduced the incidence of virus disease in vineyards planted in California since 1970. The registration and certification program, called the California Clean Stock Program, is based on research conducted by the Agricultural Research Service of the U.S. Department of Agriculture and the University of California. The objectives of the program were to test new cultivars from breeding programs and selections of standard cultivars growing in California and other parts of the United States for freedom from virus diseases and to test foreign cultivars for freedom from serious virus diseases before releasing them from quarantine.

Vines that are developed, selected, or imported and that are identified by the Department of Viticulture at the University of California are indexed for freedom from important virus diseases by the Department of Plant Pathology (Fig. 30). Vines that are free of the major virus diseases are then planted in an isolated vineyard and maintained by the Foundation Plant Materials Service. These indexed mother vines are registered with the California State Department of Food and Agriculture. Clean materials from the Foundation vineyard, after they have been rechecked for trueness to cultivar type, are released to cooperating California nurseries or commercial grape growers for further increase.

The California Clean Stock Program was originally set up to eliminate leafroll and fanleaf degeneration from commercial nursery stocks. As research continued and new diseases were found, the program objectives were expanded to include the elimination of corky bark and rupestris stem pitting. Although the program has been directed primarily at eliminating virus diseases, it could easily be augmented to include elimination of other diseases and pests that are spread by nursery stocks, such as crown gall and nematodes. However, more detailed research is needed to implement the inclusion of these additional diseases in the clean stock program.

(Prepared by A. C. Goheen and S. Nelson-Kluk)

Regulation of the International Trade in Grapevines for Propagation

The regulation of grapevines for propagation is based on certain concepts of plant quarantine. Governments establish quarantine laws and regulations to protect domestic agriculture against the introduction of foreign pests by human activities. Quarantines designed to exclude pests that can spread into an area by natural means are usually ineffective. Foreign pests named by governments as justifications for quarantine laws and regulations usually do not occur or are not widely distributed in the domestic agriculture. Those named pests that do have limited distribution in the domestic agriculture are often the targets of active programs of eradication or containment. Quarantines established solely to exclude pests that are widely distributed in the domestic agriculture are viewed as an unnecessary restraint of trade and may encourage smuggling by domestic importers or retaliatory quarantines by foreign governments.

The importation of grapevine propagation material is often regulated by countries, states, and provinces with significant grape culture. Grapevine propagation material is usually shipped as dormant scion wood rather than as whole plants or true seeds. The importation of dormant scion wood only may avoid the introduction of many of the leaf and root pests that attack whole plants in the foreign country of origin. True seed would not carry these or many other pests, but unfortunately true seed is desirable only for plant breeding purposes because grape cultivars do not breed true through seeds.

Visual inspection of dormant grape scion wood at a port of entry would detect most life stages of arthropods and most pathogens that are large enough to be seen in a light microscope or that cause obvious symptoms in dormant wood. However, such an inspection would not consistently detect new fungal or

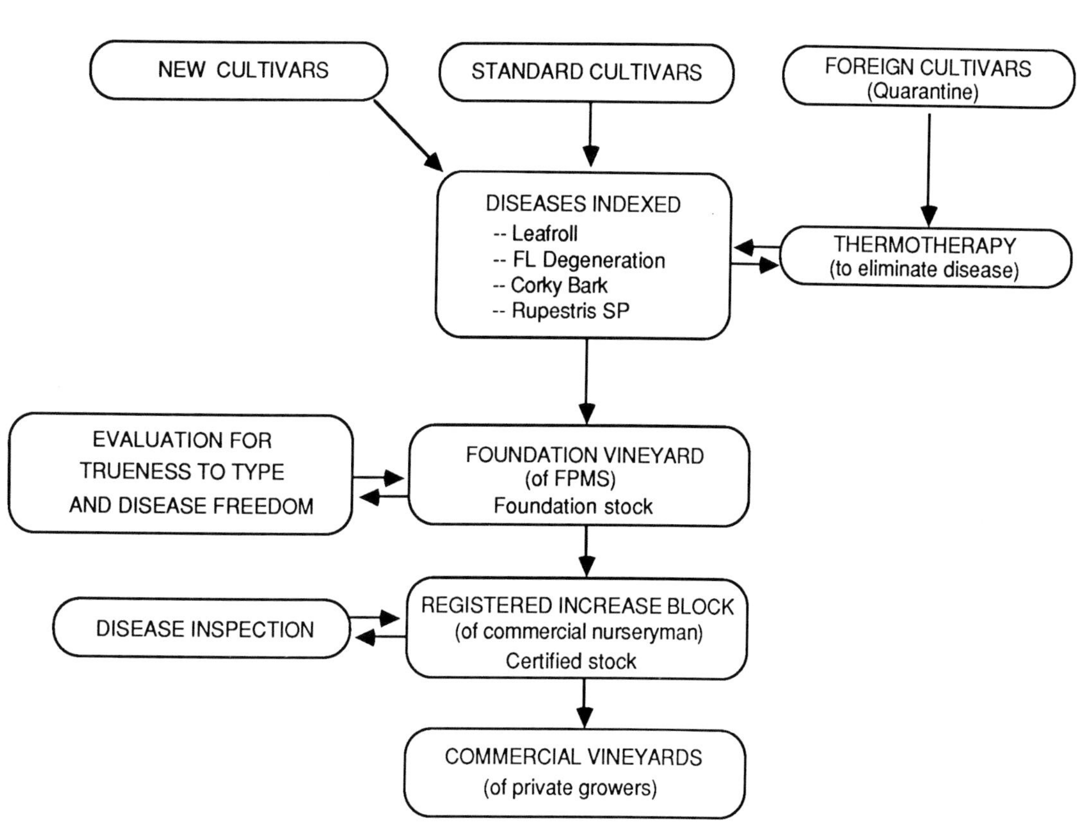

Fig. 30. Flow of grape material through the University of California clean stock program and quarantine. (Courtesy A. C. Goheen)

bacterial infections in dormant wood, single isolated arthropod eggs buried in cracks and crevices, or submicroscopic pathogens that may occur inside symptomless dormant wood. Inspections during a postentry growing period for imported grapevine propagation material would detect developing fungal and bacterial infections as well as hatching arthropods. However, specific tests are necessary to consistently detect small pathogens that may be imported in symptomless dormant wood and that may not cause obvious symptoms in the particular infected cultivar during growth. These are the pathogens that are used most often to justify prohibiting the importation of grapevine plant material for propagation. In some cases, importations of grapevines in small quantities are approved with the stipulation that tests for small internal pathogens must be made after entry.

A concern of plant protection officials in grape-growing areas is that grape material infected with a damaging pathogen that can spread by natural means, once established, would be propagated in a local vineyard. This possibility is the reason for efforts to exclude the Pierce's disease bacterium from Europe, where weeds and insects similar to the known hosts and vectors of the disease occur. Similarly, the *flavescence dorée* mycoplasmalike organism is listed in U.S. and Canadian regulations because it has not been found in North America even though its vector occurs there. Some of the important grape viruses can be transmitted to grapevines or local weeds by nematodes. Even though spread by nematode transmission is slow, these viruses can move over long distances in infected weed seeds, infested soil, or infected grapevine scion wood. Whenever exotic nematode-transmitted viruses have been discovered in a vineyard, the goal has been to destroy the infected plants before long-distance spread can occur.

Another concern shared by plant protection officials in grape-growing areas is that grapevine plant material infected with a latent but damaging pathogen might be rapidly propagated, distributed, and planted throughout local vineyards before damage caused by the pathogen became apparent. The causal agents of leafroll, corky bark, and rupestris stem pitting spread extremely slowly, if at all, in most areas. Unfortunately, these pathogens now occur in many grape-growing areas because of the propagation and planting of infected grapevines. An obvious preventive measure to minimize losses caused by these pathogens is to provide growers with planting material shown to be free of these pathogens by specific tests.

The distribution of pathogen-tested grapevine propagation material to growers is the basis of grapevine registration and certification programs that have been instituted in many grape-growing areas. These programs have eliminated or reduced to an insignificant level grape pathogens that are not transmitted to local weeds or other domestic crops under natural conditions. The success of local certification programs has convinced plant protection officials that certain pathogens that were previously widespread no longer occur or are no longer widely distributed in domestic vineyards. Consequently, a requirement that all grapevines introduced into these areas should be free of all but indigenous pathogens is a reasonable and prudent measure to protect the current health of the domestic certified grape culture. Even if a certification program has not been in place long enough to eliminate pathogens that are spread solely by plant propagation, plant protection officials can interpret the certification program as an active program of eradication aimed at eliminating these pathogens and can take regulatory action against imported grapevines infected with these pathogens.

As more vineyards in different grape-growing areas participate in local certification programs, the growers and regulatory officials in these areas are likely to develop more restrictive quarantine regulations to protect the healthier local grape culture from pathogens imported with diseased grapevines. The more restrictive quarantine regulations will probably require other grape-growing countries to develop acceptable certification programs if they wish to export their grapevines to areas where certification programs have been successful. The ultimate outcome of these certification and regulatory activities in many grape-growing areas will be freer international exchange of pathogen-tested grapevine propagation material.

Selected References

Kahn, R. P. 1977. Plant quarantine: Principles, methodology, and suggested approaches. Pages 289–307 in: Plant Health and Quarantine in International Transfer of Genetic Resources. W. B. Hewitt and L. Chiarappa, eds. CRC Press, Cleveland. 346 pp.

Martelli, G. P. 1978. Nematode-borne viruses of grapevine, their epidemiology and control. Nematol. Mediterr. 6:1–27.

Nielson, M. W. 1968. The leafhopper vectors of phytopathogenic viruses (Homoptera: Cicadellidae): Taxonomy, biology, and virus transmission. U.S. Dep. Agric. Tech. Bull. 1382.

Raju, B. C., Goheen, A. C., and Frazier, N. W. 1983. Occurrence of Pierce's disease bacteria in plants and vectors in California. Phytopathology 73:1309–1313.

Stobbs, L. W., and Van Schagen, J. G. 1984. Occurrence of tomato black ring virus on grapevine in southern Ontario. Can. Plant Dis. Surv. 64:3–5.

Uyemoto, J. K., Taschenberg, E. F., and Hummer, D. K. 1977. Isolation and identification of a strain of grapevine Bulgarian latent virus in Concord grapevine in New York State. Plant Dis. Rep. 61:949–953.

(Prepared by J. A. Foster)

Appendix. Equivalent Names of Grape Diseases and Disorders

Common Name Causal Factors	Other Names	French	German	Italian	Spanish
Acid soil	Sour soil	Sol acide	Saurer Boden Bodensäure	Aciditá del terreno	Suelo ácido
Air pollution injury		Pollution atmosphérique Pollution aérienne	Schäden durch Luftverunreinigung	Danni da inquinamento dell'aria	Daño por contaminación del aire
Hydrogen fluoride		Acide fluorhydrique, gaz fluoré	Fluorwasserstoff	Acido fluoridrico	Acido fluorhídrico
Ozone	Oxidant stipple	Ozone	Ozon	Ozono	Ozono
Sulfur dioxide		Dioxide de soufre	Schwefeldioxid	Anidride solforosa	Bióxido de azufre
Alkali injury Excess sodium		Excès de sodium	Alkalischäden	Danni da alcalinità	Daño por sales
Alternaria rot *Alternaria alternata*		Alternariose	Alternariafäule	Alternariosi	Pudrición por *Alternaria*
Angular leaf spot *Mycosphaerella angulata*				Maculatura fogliare angolare	Mancha angular de hojas
Anthracnose *Elsinoë ampelina*	Bird's-eye rot	Anthracnose	Anthraknose Schwarzer Brenner	Antracnosi	Antracnosis
Armillaria root rot *Armillaria mellea*	Shoestring root rot Mushroom root rot Oak root fungus	Pourridié à armillaire Pourridié agaric	Wurzelfäule durch Hallimasch Heckenschwamm Honigpilz	Marciume radicale fibroso	Pudrición de raíces por *Armillaria*
Asteroid mosaic Undetermined, viruslike		Mosaïque étoilée	Sternmosaik der Rebe	Mosaico stellare	Mosaico asteroide
Bacterial blight *Xanthomonas ampelina*	Bacterial necrosis	Nécrose bactérienne Maladie d'Oléron	Bakteriennekrose	Mal nero Batteriosi	Tizón Necrosis bacteriana
Bitter rot *Greeneria uvicola* (*Melanconium fuligineum*)		Pourriture amère	Bitterfäule		Pudrición amarga
Black dead arm *Botryosphaeria stevensii*				Necrosi del legno	Brazo muerto negro
Black measles Presumably toxins from wood-rotting fungi	Spanish measles	Esca Apoplexie		Mal dell'esca	Sarampión negro
Black rot *Guignardia bidwellii*		Black-rot Pourriture maculée	Schwarzfäule Schwarze Trockenfäule der Beeren	Marciume nero degli acini	Pudrición negra

(*continued on next page*)

Common Name Causal Factors	Other Names	French	German	Italian	Spanish
Bois noir Mycoplasmalike organism	Black wood disease	Bois noir	Vergilbungskrankheit Schwarzholzkrankheit	Legno nero	Madera negra
Blue mold *Penicillium* spp.	Penicillium rot	Pourriture amère Pourriture acide (sur grappe)	Grünfäule Speckfäule durch *Penicillium glaucum* und *Penicillium* spp.	Marciume azzurro	Pudrición por hongo azul
Botrytis bunch rot and blight *Botrytis cinerea*	Gray mold	Pourriture grise	Botrytis-Traubenfäule und Botrytis-Triebfäule Graufäule	Muffa grigia	Pudrición gris de racimos y tizón de cañas
Cercospora leaf spot *Phaeoramularia dissiliens*		Cercosporiose		Cercosporiosi	Mancha foliar por *Cercospora*
Citrus nematode *Tylenchulus semipenetrans*		Nématode des citrus Nématode du dépérissement des agrumes	Citrusnematode	Nematode degli agrumi	Nemátodo de los cítricos
Cladosporium leaf spot *Cladosporium viticola*		Cladosporiose			Mancha foliar por *Cladosporium*
Cladosporium rot *Cladosporium herbarum*		Cladosporiose			Pudrición por *Cladosporium*
Corky bark Undetermined, viruslike	Grapevine stem pitting	Ecorce liégeuse Maladie de l'écorce liégeuse	Korkrindenkrankheit der Rebe	Suberosi corticale	Corteza corchosa
Crown gall *Agrobacterium tumefaciens*		Broussin parasitaire	Mauke Grind Krebs des Weinstocks	Tumore batterico	Agalla de la corona
Dagger nematodes *Xiphinema* spp.		Nématode poignard Nématode à dague	Stilettälchen	Nematode daghiforme	Nemátodo daga
Dematophora root rot *Rosellinia necatrix*	Western white root rot	Pourridié laineux	Wurzelschimmel Wurzelfäule	Marciume radicale bianco	Pudrición blanca de raíces
Diplodia cane dieback and bunch rot *Diplodia natalensis*	Diplodia dieback Cane-tip blight			Necrosi basipeta da *Diplodia*	Tizón apical de sarmientos
Downy mildew *Plasmopara viticola*		Mildiou	Peronospora Falscher Mehltau Blattfallkrankheit	Peronospora	Mildiú
Drought stress		Sécheresse Coup de soif	Dürreschäden Trocken(heits)schäden	Scompensi idrici	Tensión hídrica
Enation Undetermined, viruslike		Maladie des énations	Enationenkrankheit der Rebe	Malattia delle enazioni	Enaciones
Esca Presumably toxins from wood-rotting fungi	Apoplexy	Esca Apoplexie	Apoplexie Schlaganfall der Rebe	Mal dell'esca Apoplessia	Yesca
Eutypa dieback *Eutypa lata*		Eutypiose	Eutypose Holzhartfäule	Eutipiosi	Eutipiosis Brazo muerto
Excess water injury		Asphyxie Excès d'eau	Schäden durch Wasserüberschuss Stauende Nässe	Danni da eccessi idrici	Daño por exceso de agua
Fanleaf degeneration Grapevine fanleaf virus	Infectious degeneration and decline	Court-noué Dégénérescence infectieuse	Reisigkrankheit der Rebe	Arricciamento Degenerazione infettiva	Hoja de abanico Virus del entrenudo corto infeccioso de la vid

Common Name Causal Factors	Other Names	French	German	Italian	Spanish
Fasciation Genetic disorder		Fasciation	Verbänderung Fasziation	Fasciazioni	Fasciación
Flavescence dorée Mycoplasmalike organism		Flavescence dorée Maladie du Baco 22A	Goldgelbe Vergilbung der Rebe	Flavescenza dorata	Flavescencia dorada
Fleck Undetermined, viruslike		Marbrure	Marmorierung	Maculatura infettiva	Punteado
Grape root rot *Roesleria subterranea*	Roesleria root rot	Pourridié morille	Roesleria-Wurzelfäule	Marciume radicale da *Roesleria*	Pudrición de raíces por *Roesleria*
Grapevine yellows Mycoplasmalike organisms		Panachure Jaunisse de la vigne	Vergilbungen	Giallumi della vite	Amarillamientos de la vid
Hail injury		Blessures de grêle Dégât de la grêle	Hagelschäden	Danni da grandine	Daño por granizo
Heat stress	High-temperature injury Almeria spot	Coup de chaleur Coup de soleil Coup de pouce Echaudage	Hitzeschäden	Colpo di calore	Tensión por calor
Lance nematodes *Hoplolaimus* spp. *Rotylenchus* spp.				Nematode lanciforme	Nemátodo lanza
Leaf blight *Mycosphaerella personata*	Isariopsis leaf spot *Pseudocercospora vitis*			Seccume delle foglie	Tizón foliar
Leaf blotch *Briosia ampelophaga*				Chiazzatura fogliare	Roncha foliar
Leafroll Undetermined, viruslike		Enroulement	Blattrollkrankheit der Rebe	Accartocciamento fogliare	Enrollamiento de la hoja
Lesion nematodes *Pratylenchus* spp.		Nématode des lésions racinaires	Läsionenälchen	Nematode delle lesioni	Nemátodo de las lesiones
Lightning injury		Dégâts de foudre Coup de foudre	Blitzschäden	Danni da fulmine	Daño por relámpagos
Macrophoma rot *Botryosphaeria dothidea*	Botryosphaeria rot and necrosis			Marciume da *Macrophoma*	Pudrición por *Botryosphaeria*
Needle nematodes *Longidorus* spp.		Nématode aiguille	Nadelälchen	Nematode dell'ago	Nemátodo de aguja
Peach rosette mosaic virus decline Peach rosette mosaic virus	Grapevine degeneration Grapevine decline Berry shelling disease	Mosaïque à rosettes du pêcher	Rosettenmosaik des Pfirsichs	Mosaico con rosettamento del pesco	Mosaico rosetado del durazno
Phomopsis cane and leaf spot *Phomopsis viticola*		Excoriose	Schwarzflecken-krankheit	Escoriosi Necrosi corticale	Mancha foliar y de sarmientos
Phymatotrichum root rot *Phymatotrichum omnivorum*	Cotton root rot Ozonium root rot Texas root rot			Marciume radicale da *Phymatotrichum*	Pudrición Texana
Phytophthora crown and root rot *Phytophthora* spp.	Root and collar rot	Pourriture du collet	Wurzelfäule Kragenfäule	Marciumi radicali e del colletto	Pudrición del cuello
Pierce's disease *Xylella fastidiosa*		Maladie de Pierce	Pierce'sche Krankheit	Malattia di Pierce	Enfermedad de Pierce

(continued on next page)

Common Name Causal Factors	Other Names	French	German	Italian	Spanish
Pin nematodes *Paratylenchus* spp.		Nématode épingle	Pinnematode	Nematode spilliforme	Nemátodo alfiler
Powdery mildew *Uncinula necator*	Oidium	Oïdium	Oidium Mehltau Schimmel Äscherich	Oidio Mal bianco	Cenicilla Oidio
Reniform nematode *Rotylenchulus* spp.		Nématode réniforme		Nematode reniforme	Nemátodo reniforme
Rhizopus rot *Rhizopus* spp.		Pourriture à *Rhizopus* Pourriture acide		Marciume lanoso	Pudrición por *Rhizopus*
Ring nematode *Criconemella xenoplax*			Ringnematode	Nematode ad anelli	Nemátodo anillado
Ripe rot *Colletotrichum gloeosporioides*					Antracnosis
Root-knot nematodes *Meloidogyne* spp.		Nématode à galles Nématode cécidogène Nématode des racines noueuses	Wurzelgallenälchen	Nematode galligeno	Nemátodo agallador Nemátodo nodulador
Rotbrenner *Pseudopezicula tracheiphila*		Rougeot parasitaire Brenner	Roter Brenner Sang Brand	Rossore delle foglie	Quemadura roja
Rupestris speckle Physiological disorder		Moucheture du rupestris		Picchettatura della *Vitis rupestris*	Manchado rupestris
Rupestris stem pitting					
Rust *Physopella ampelopsidis*		Rouille de la vigne	Rost	Ruggine	Roya
Salt toxicity	Salinity injury	Toxicité saline Effets toxiques d'excès de sels solubles	Salztoxizität Salzschäden	Danni da salinità	Daño por sales
Septoria leaf spot *Septoria ampelina*	Melanose	Mélanose infectieuse	Melanose	Melanosi	Mancha foliar por *Septoria*
Shoot necrosis Undetermined, viruslike		Nécrose des sarments de la vigne	Triebnekrose der Rebe	Necrosi dei germogli	Necrosis del tallo
Sour bunch rots Various yeasts and acetic acid bacteria		Pourriture acide Diverses levures et bactéries acétiques	Sauerfäule Verschiedene Hefen und Essigsäure-bakterien	Marciume acido Lieviti e batteri acetici diversi	Pudrición ácida Varias levaduras y bacterias
Spiral nematodes *Helicotylenchus* spp.		Nématode spiralé	Spiralälchen	Nematode a spirale	Nemátodo espiral
Spring freeze injury	Frost	Dégâts de froid au printemps Gelées printanières Gel tardif Gel de printemps	Schäden durch Frühjahrsfrost Spätfrostschäden	Danni da gelo primaverile	Daño por heladas tardías
Stem necrosis Physiological disorder	Grape peduncle necrosis Stalk necrosis Water berry	Dessèchement de la rafle Flétrissement pédonculaire	Stiellähme	Disseccamento del rachide	Necrosis del raquis
Stubby-root nematode *Paratrichodorus christiei*		Rattle Nématode des racines tronquées	Borstenwurzelälchen	Nematode delle radici tronche	Nemátodo atrofiador de raíces

Common Name Causal Factors	Other Names	French	German	Italian	Spanish
Stunt nematodes *Tylenchorhynchus* spp.		Nématode du rabougrissement			
Tar spot *Rhytisma vitis*				Croste nere	Mancha de chapopote
Tobacco ringspot virus decline Tobacco ringspot virus		Taches annulaires du tabac	Tabakringflecken	Maculatura anulare del tabacco	Mancha anular del tabaco
Tomato ringspot virus decline Tomato ringspot virus	Grape yellow vein Little berry	Taches annulaires de la tomate	Tomatenringflecken	Maculatura anulare del pomodoro	Mancha anular del tomate
Vein mosaic Undetermined, viruslike	Veinclearing	Mosaïque des nervures	Adernmosaik	Mosaico delle nervature	Mosaico de las nervaduras
Verticillium wilt *Verticillium dahliae*		Flétrissement à *Verticillium* Verticilliose	Verticillium-Welke	Verticilliosi	Marchitamiento por *Verticillium*
White rot *Coniella diplodiella*		Rot blanc Coître	Weissfäule	Marciume bianco degli acini Carie bianca	Pudrición blanca por *Coniella*
Wind and sand injury		Dégâts de vent et de sable Folletage	Windschäden Schäden durch Sandwehen	Danni da vento e sabbia	Daño por viento y arena
Winter injury		Dégâts de gel d'hiver Gélivure	Winter(frost)schäden Schäden durch tiefe Temperaturen	Danni da freddo	Daño por frío
Yellow speckle Undetermined, viruslike		Moucheture jaune	Gelbsprenkelung	Picchettatura gialla	Punteado amarillo
Zonate leaf spot *Cristulariella moricola*	Target spot			Maculatura zonata	Mancha zonada

Glossary

C—Celsius or centigrade (°C = (°F − 32) × 5/9)
cm—centimeter (1 cm = 0.01 m = 0.3937 in.)
F—Fahrenheit (°F = (°C × 9/5) + 32)
g—gram (1 g = 0.0353 oz; 453.6 g = 1 pound)
ha—hectare (1 ha = 10,000 m^2 = 2.471 acres)
hr—hour
kg—kilogram (1 kg = 1,000 g = 2.205 pounds)
km—kilometer (1 km = 1,000 m = 0.6214 mi)
L—liter (1 L = 1.057 quarts liquid [U.S.])
m—meter (1 m = 39.37 in.)
meq—milliequivalent
min—minute
ml—milliliter (1 ml = 0.001 L)
mm—millimeter (1 mm = 0.001 m)
μm—micrometer (1 μm = 10^{-6} m)
nm—nanometer (1 nm = 10^{-9} m)
ppm—parts per million

abaxial—directed away from the stem of a plant; pertaining to the lower surface of a leaf
abscise (n. abscission)—to separate from a plant, as leaves or berries do
acervulus (pl. acervuli)—saucer-shaped or cushionlike fungal fruiting body bearing conidiophores, conidia, and sometimes setae
acicular—slender and pointed; needle-shaped
acid soil—soil with an acidic reaction (less than pH 7)
acropetal—upward from the base to the apex of a shoot or cane
adaxial—directed toward the stem of a plant; pertaining to the upper surface of a leaf
adventitious—arising from other than the usual place, as roots from a stem rather than as branches of a root
alkaline—having basic (nonacidic) properties
allantoid—slightly curved with rounded ends; sausagelike in form
ampelographic—relating to the description of grape cultivars
amphigenous—making growth all around or on two sides
amphigynous—having the antheridium surrounding and penetrated by the oogonial stalk (*see* oogonium)
anamorph—the asexual form (also called the imperfect state) in the life cycle of a fungus, when asexual spores (such as conidia) or no spores are produced
anastomosis (pl. anastomoses)—fusion, as of hyphal strands, and combination of their contents
anther—the pollen-bearing portion of a stamen
antheridium (pl. antheridia)—male sexual organ found in some fungi
anthesis—the time of pollination or bloom
anthocyanin—blue, purple, red, or pink water-soluble flavanoid pigment in cell sap
anthracnose—disease caused by acervuli-forming fungi (order Melanconiales) and characterized by sunken lesions and necrosis
antibody—a specific protein formed in the blood of warm-blooded animals in response to the injection of an antigen
antigen—any foreign chemical (normally a protein) that induces antibody formation in animals
apex (pl. apexes, adj. apical)—tip of a root or shoot, containing the apical meristem
apical dominance—inhibition of growth of lateral buds or branches by the topmost growing point
apothecium (pl. apothecia)—open, cuplike or saucerlike, ascus-bearing fungal fruiting body
appressorium (pl. appressoria)—swollen, flattened portion of a fungal filament that adheres to the surface of a higher plant, thus providing anchorage for invasion by the fungus
arm—in viticultural usage, a branch more than one year old that is attached to a trunk and that bears canes or spurs
ascocarp—sexual fruiting body (ascus-bearing organ) of an ascomycete
ascogenous—developing or originating from an ascus
ascomycete—member of a class of fungi that produce sexual spores (ascospores) endogenously within an ascus
ascospore—sexual spore borne in an ascus
ascus (pl. asci)—saclike cell in which ascospores (typically eight) are produced
asexual—vegetative; without sex organs, sex cells, or sexual spores, as the anamorph of a fungus
axil—the angle formed by a leaf petiole and the stem
axillary bud—bud that develops in the axil of a leaf (also called a lateral bud; *see* lateral shoot)

bacilliform—shaped like a blunt, thick rod
bark—in viticultural usage, brown periderm (a protective tissue) or, more generally, all the tissues outside the cambium (including the phloem and periderm)
basidiomycete—member of a class of fungi that form sexual spores (basidiospores) on a basidium
basidiospore—haploid spore of a basidiomycete
basidium (pl. basidia, adj. basidial)—short, club-shaped fungus cell on which basidiospores are produced
basipetal—downward from the apex toward the base of a shoot or cane; referring to development in the direction of the base so that the apical part is oldest
berm—ridge of soil in a vine row
bicellular—two-celled
biflagellate—having two flagella
biological control—disease or pest control through counterbalance by microorganisms and other natural components of the environment
biovar—infrasubspecific group of organisms differentiated from other such groups within the same species by biochemical or physiological properties (also called biotype)
blade—the flat, expanded portion (lamina) of a leaf
blight—any sudden, severe, and extensive spotting, discoloration, wilting, or destruction of leaves, flowers, stems, or entire plants, usually attacking young, growing tissues (in disease names, often coupled with the name of the affected part of the host, e.g., leaf blight, blossom blight, shoot blight)
bloom—flowering, as indicated by the shedding of caps; the waxy coating on grape berries, which gives a frosted appearance to dark-colored varieties
botryose—shaped like a bunch of grapes
brush—in viticultural usage, the broken ends of the vascular bundles that remain attached to the pedicel when a mature berry is pulled off
bud—*see* compound winter bud
budbreak—stage of bud development when green tissue becomes visible

calcareous—rich in calcium carbonate (lime)
callose—carbohydrate deposit on sieve plates of sieve elements of phloem, indicating dormancy, injury, etc.
callus—parenchyma tissue that grows over a wound or graft and protects it against drying or other injury
calyptra—*see* cap
calyx—outermost whorl of organs of a flower (very small in a grape flower)
cambium (pl. cambia)—sheath of meristematic cells in stem and root, which divide primarily tangentially, producing secondary xylem toward the inside and secondary phloem toward the outside
cane—in viticultural usage, a mature woody shoot from leaf fall through its second year
canker—necrotic, localized diseased area
canopy—in viticultural usage, the mass of leaf-bearing shoots, measured in height, width, or distribution
cap—calyptra, or corolla, of a grape flower, formed by petals

interlocking along their edges and at anthesis separating along their bases, thus falling as a unit

cap stem—*see* pedicel

carbohydrate—any of various chemical compounds of carbon, hydrogen, and oxygen, such as sugars, starches, and cellulose

causal agent—organism or agent that produces a given disease

cellulose—a carbohydrate constituting the primary substance of cell walls

centrum—the structures within an ascocarp

chalaza—point of attachment of a seed to the ovary

chimera—a plant with several tissue sectors or layers differing in genetic or chromosomal constitution from the original plant

chlamydospore—thick-walled or double-walled asexual resting spore formed by modification of a hyphal segment

chlorophyll—green pigment of plants that absorbs light energy and makes it effective in photosynthesis

chloroplast—disklike structure containing chlorophyll in a plant cell

chlorosis (adj. chlorotic)—abnormal plant color of light green or yellow due to incomplete formation or destruction of chlorophyll

cirrus (pl. cirri)—a curllike tuft; a tendrillike mass or "spore horn" of forced-out spores

clavate—club-shaped

cleistothecium (pl. cleistothecia)—closed, usually spherical, ascus-containing structure of powdery mildew fungi

clone—vegetatively (asexually) propagated plant or member of a group of such plants derived from a single original plant

closterovirus—a long (600–2,000 nm), thin, very flexuous, threadlike plant virus

coenocytic—multinucleate (e.g., pertaining to a multinucleate plant body enclosed within a common wall or a fungus filament lacking cross-walls)

collarette—a cup-shaped structure at the apex of a phialide

compound winter bud—in viticultural usage, an overwintering bud at the node of a cane (also called an eye)

concentric—pertaining to circles with a common center but different diameters

conidiogenous—producing conidia

conidioma (pl. conidiomata)—specialized multihyphal conidia-bearing structure (*see*, e.g., acervulus, pycnidium, synnema)

conidiophore—specialized fungal hypha on which conidia (conidiospores) are produced

conidium (pl. conidia)—asexual spore formed by abstriction and detachment of part of a hyphal cell at the end of a conidiophore and germinating by a germ tube

cordon—in viticultural usage, an extension of a trunk, trained along a horizontal wire, and capable of bearing arms, spurs, and canes

cortex (adj. cortical)—region of parenchyma tissue between the epidermis and the phloem in stems and roots

coulure—excessive drop (shatter) of flowers

crawler—first-instar nymph of grape phylloxera, scale insects, and related groups

cross protection—a phenomenon whereby plants infected with one strain of a virus do not develop additional symptoms when inoculated with a second strain of the same virus

crown—point where a main stem (trunk) and root join, at or just below the soil surface

cultivar (abbr. cv.)—a cultivated plant variety, or cultural selection

culture—artificial growth and propagation of organisms on nutrient media or living plants

cupulate—cuplike in form

cuticle (adj. cuticular)—water-repellent waxy covering (cutin) of epidermal cells of plant parts such as leaves, stems, and fruits; the outer sheath or membrane of nematodes

cutin—*see* cuticle

cutting—in grapevine propagation, a piece of shoot (softwood cutting) or cane (hardwood cutting) induced to produce adventitious roots

cymbiform—boat-shaped

cytoplasm—inner substance of a cell exclusive of the nucleus

degree Brix—unit of measurement on the Brix scale (the Brix value represents the total soluble solid content of grapes, or approximately the percentage of grape sugars in the juice)

dehiscence—opening when mature, by pores or by breaking into parts

deliquesce—to become liquid after maturing

denticulation—small, toothlike projection

diagnostic—distinctive, as of a distinguishing characteristic serving to identify or determine the presence of a disease or other condition

diapause—a period of arrested development or suspended animation

diaphragm—*see* pith

dichotomous—branching, often successively, into two more or less equal arms

dieback (v. die back)—progressive death of shoots, leaves, or roots, beginning at the tips

differentiation—the physiological and morphological changes that occur in a cell, tissue, or organ during development from a juvenile state to a mature state

dioecious—having male (staminate) flowers and female (pistillate) flowers on separate plants

diploid—having a double set of chromosomes ($2n$ chromosomes) per cell

direct producers—vines obtained by crossing *Vitis vinifera* vines with American *Vitis* spp., which combine to some extent the quality of *V. vinifera* fruit and the resistance of the American vines to fungus diseases and insect pests, particularly phylloxera

disbudding—in viticultural usage, the removal of unwanted buds from a cutting or graft

discomycete—ascomycetous fungus that generally bears asci on apothecia

dissemination—spread of infectious material (inoculum) from diseased to healthy plants

distal—far from the point of attachment or origin, as opposed to proximal

dormancy—nongrowing condition of a plant, caused by internal factors (as in endodormancy) or environmental factors (as in ectodormancy)

drop—shatter: dropping of unpollinated flowers and young fruits from a grape cluster about a week after bloom

echinulated—having spines or other sharp projections

ectoparasite—parasite living outside its host

ELISA—*see* enzyme-linked immunosorbent assay

enation—epidermal outgrowth

endemic—native to or peculiar to a locality or region

endocellular—inside the cell

endoparasite—parasite living within its host

enzyme—protein that catalyzes a specific biochemical reaction

enzyme-linked immunosorbent assay—a serological test in which the sensitivity of the antibody-antigen reaction is increased by attaching an enzyme to one of the two reactants

epidemic—general and serious outbreak of disease (used loosely of plants)

epidemiology (adj. epidemiologic)—the study of factors influencing the initiation, development, and spread of infectious disease

epidermis (adj. epidermal)—outermost layer of cells on plant parts

epinasty—downward curvature of a leaf, leaf part, or stem

eradicant—chemical used to eliminate a pathogen from a host or an environment

eradicate—to destroy or remove a pest or pathogen after disease has become established

eriophyid—describing mites belonging to the family Eriophyidae, commonly known as gall, rust, bud, and blister mites

erumpent—breaking out or erupting through the surface

evapotranspiration—total amount of water removed from the vineyard ecosystem by evaporation and by transpiration

explant line—descendants growing artificially from living tissues outside their normal habitat

extrude—to push out; emit to the outside

exudate—substance that is excreted or discharged; ooze

eye—*see* compound winter bud

facultative—capable of changing life-style, e.g., from saprophytic to parasitic or the reverse

fallow—cultivated land kept free from a crop or weeds during the normal growing season

fasciation—malformation in canes, shoots, or floral organs manifested as enlargements and flattening and sometimes curving as if several parts were fused

fastidious—referring to prokaryotic organisms with special growth and nutritional requirements

female—having flowers with pistils but no stamens

fiber—elongated, narrow, thick-walled cell in xylem or phloem

filament (adj. filamentous)—thin, flexible, threadlike structure; the stalklike, anther-bearing portion of a stamen

filiform—long, needlelike

flaccid—wilted, lacking turgor

flagellum—hairlike or whiplike appendage of bacterial cells or fungus zoospores, providing locomotion

frass—excrement of an insect

fructification—fruiting body

fruit—mature ovary (berry) or cluster of mature ovaries

fruiting body—any of various complex, spore-bearing fungal structures

fruit set—stage of berry development (after pollination and fertilization) one to three weeks after blossoming, when most flowers

have fallen and remaining ovaries start to enlarge and develop into berries (also called berry set stage or shatter stage)
fumigant—vapor-active chemical used in the gaseous phase to kill or inhibit the growth of microorganisms or other pests
fungicide (adj. fungicidal)—chemical or physical agent that kills or inhibits the growth of fungi
fusiform—narrowing toward the ends

gall—outgrowth or swelling of unorganized plant cells produced as a result of attack by bacteria, fungi, or other organisms
gelatinous—resembling gelatin or jelly
genetic—relating to heredity; referring to heritable characteristics
genome—set or group of chromosomes
genotype—genetic constitution of an individual, in contrast to its appearance, or phenotype
genus (pl. genera)—group of related species
germinate—to begin growth of a seed or spore
germ tube—initial hyphal strand from a germinating fungus spore
giant cells—multinucleate cells formed by disintegration of cell walls (also called syncytia, in nematode infections)
girdle—to circle and cut through; to destroy vascular tissue as in a canker or knife cut that encircles the stem
girdling—removal of a complete ring of bark from a shoot, cane, or trunk (also called ringing), temporarily interrupting the downward translocation of metabolites
graft—in grapevine propagation, a scion (stem piece of a cultivar with a bud or buds) inserted into a rootstock (stem piece induced to produce adventitious roots) so that their cambia unite
gram-negative, gram-positive—pertaining to bacteria that release or retain, respectively, the violet dye in Gram's stain
guttation—exudation of watery, sticky liquid from stomata or hydathodes of leaves
guttule (adj. guttulate)—oillike drop inside spores

haploid—having a single complete set of chromosomes (*n* chromosomes)
hardwood cutting—in grapevine propagation, a piece of cane with buds induced to produce adventitious roots
haustorium (pl. haustoria)—specialized outgrowth (of a stem, root, or mycelium) that penetrates a host plant and extracts nutrients
head—in viticultural usage, upper portion of a vine consisting of trunk and arms
herbaceous—nonwoody, as a plant or plant part
herbicide—chemical that kills plants (also applied to chemicals that limit the growth of such plants)
hermaphroditic—perfect: having stamens and pistil (or pistils) in the same flower
heterokaryotic—referring to the condition of having two or more genetically different nuclei in the same protoplast
heterothallic—pertaining to species of fungi in which the sexes are separated in different mycelia
homothallic—pertaining to species of fungi in which both sexes are present in the same mycelium, so that sexual reproduction can occur without interaction with another mycelium
hyaline—colorless, transparent
hybrid (v. hybridize)—sexually produced offspring of genetically differing parents (if the parents are of different species, the offspring is an interspecific hybrid; in grapes, further vegetative propagation continues it as a clone)
hydathode—epidermal leaf structure specialized for secretion or exudation of water
hyperplasia—abnormal increase in the number of cells in a tissue or organ, resulting in formation of galls or tumors
hypersensitive—extremely or excessively sensitive; having a type of resistance resulting from extreme sensitivity to a disease
hypertrophy—abnormal increase in the size of cells in a tissue or organ, resulting in formation of galls or tumors
hypha (pl. hyphae, adj. hyphal)—tubular filament of a fungus
hypophyllous—on the undersurface of a leaf

immunofluorescence—techniques involving the visualization of the antibody-antigen reaction in the light microscope through the use of conjugated fluorescent probes
immunosorbent electron microscopy—techniques involving the visualization of the antibody-antigen reaction in the electron microscope
imperfect state—*see* anamorph
incipient—early in development (of a disease or condition)
incubation period—time between infection by a pathogen and appearance of symptoms
indexing—determination of the presence of disease in a grapevine by removing buds or other parts for inoculation of a susceptible indicator plant that exhibits specific symptoms of a transmissible disease
indicator plant—plant that reacts to a pathogen (such as a virus) or to an environmental factor with specific symptoms, used to identify pathogens or determine the effects of environmental factors
infection—process in which a pathogen enters, invades, or penetrates and establishes a parasitic relationship with a host plant
infection court—site in or on a host plant where infection can occur
infectious—capable of spreading disease from plant to plant
infective—referring to an organism or virus able to attack a host and cause infection; referring to a vector carrying or containing a pathogen and able to transfer it to a host plant, causing infection
infest—to attack as a parasite (used especially of insects and nematodes); to contaminate, as with microorganisms; to be present in numbers
inflorescence—cluster of flowers
inoculate (n. inoculation)—to place inoculum in an infection court
inoculum—pathogen or pathogen part (e.g., spores, mycelium) that infects plants
inoperculate—referring to the opening of an ascus or sporangium by an irregular apical split to discharge the spores
intercellular—between cells
internode—the portion of a stem between two adjacent nodes
interspecific hybrid—*see* hybrid
interveinal—between (leaf) veins
intracellular—within cells
intraspecific—within a species
in vitro—in glass, on artificial media, or in an artificial environment
isolate—pure microbial culture, separated from its natural origin

Koch's postulates—proof of pathogenicity by consistent association of a pathogen with a disease, isolation of the pathogen in pure culture, reinoculation of the host with the pathogen resulting in the same symptoms or disease originally observed, and reisolation of the pathogen (identical to the original isolate) from the newly inoculated plant

lageniform—swollen at the base; narrowed at the top
lamina (pl. laminae)—the broad, expanded portion, or blade, of a leaf
larva (pl. larvae)—juvenile stage of certain animals (e.g., nematodes, aphids) between the embryo and the adult
latent—present but not manifested or visible, as a symptomless infection
latent bud—in viticultural usage, a nongrowing bud, originally axillary, usually located on wood more than one year old, capable of developing into a shoot
lateral bud—bud formed in the axil of a leaf
lateral shoot—in viticultural usage, a shoot (also called a summer lateral) produced from a lateral or axillary bud of a leaf and developing in the same season as the bud
layer—long cane from a vine induced to root to replace a missing adjacent vine
leaf scar—scar left on a cane after leaf fall
lenticel (adj. lenticular)—a group of loosely arranged, corky cells (as on the epidermis of a mature grape berry, pedicel, or stem) permitting gas exchange
lesion—wound or delimited diseased area
lobe—a rounded projection of an organ
locule (adj. locular)—cavity, especially in a fungus stroma
lumen—the central cavity of a cell or other structure

macroconidium (pl. macroconidia)—the larger and generally more diagnostic type of conidium produced by a fungus that also has microconidia
macrocyclic—referring to a long-cycled rust producing at least one type of binucleate spore in addition to the teliospore
male—having flowers with stamens but no pistils
male-sterile—pertaining to plants that produce no viable pollen
mating type—a strain of organisms incapable of sexual reproduction with one another but capable of such reproduction with members of other strains of the same organism
matrix—the material in which an organism or organ is embedded
mechanical injury—injury of a plant part by abrasion, mutilation, or wounding
meristem (adj. meristematic)—plant tissue characterized by frequent cell division, producing cells that become differentiated into specialized tissues
mesophyll—central, internal, nonvascular tissue of a leaf, consisting of

the palisade and spongy mesophyll
microconidium (pl. microconidia)—the smaller type of conidium produced by a fungus that also has macroconidia
microfissure—microscopic narrow opening
micrografting—grafting of shoot tips or plants growing in vitro
micronutrient—nutrient required in small quantities (trace amounts) for normal growth and reproduction but toxic at high concentrations
microsclerotium (pl. microsclerotia)—microscopic, dense aggregate of darkly pigmented, thick-walled hyphal cells
millerandage—a disorder characterized by excessive production of small, seedless, green grape berries, called shot berries
MLO—*see* mycoplasmalike organism
molt—to shed a cuticle or body encasement during a growth phase
monophialidic—having one phialide
morphology—the study of the form of organisms
mosaic—disease symptom characterized by nonuniform foliage coloration, with a more or less distinct intermingling of normal and light green or yellowish patches, usually caused by a virus; mottle
mottle—disease symptom characterized by light and dark areas in an irregular pattern on a leaf or fruit
mucilaginous—viscous, slimy
multiseptate—having many septa
mummify—to dry and shrivel up
mummy—a dried and shriveled grape berry
muscadine—popular name for cultivars of *Vitis rotundifolia*
mutation—heritable genetic change in a cell or plant
mycelium (pl. mycelia)—mass of hyphae constituting the body (thallus) of a fungus
mycoparasite—fungus parasitic on another fungus
mycoplasmalike organisms (MLOs)—microorganisms found in phloem tissue that resemble mycoplasmas in all respects except that they cannot yet be grown on artificial nutrient media

necrosis (adj. necrotic)—death of tissue, usually accompanied by black or brown darkening
nectary—one of the five swellings from the base of the ovary of a grape flower, the function of which is unknown
nematicide—agent, usually a chemical, that kills or inhibits nematodes
nepovirus—any of a group of nematode-transmitted polyhedral viruses
node—enlarged portion of a shoot or cane at which leaves, clusters, tendrils, or buds are located
nodosity—gall, usually bent and club-shaped, formed on feeder roots by phylloxera
nonseptate—describing fungus filaments without cross-walls
nymph—juvenile stage of an insect, with incomplete metamorphosis but superficially resembling the adult

obligate parasite—organism that can grow or reproduce only on or in living tissue
obovate—inversely ovate
oogonium (pl. oogonia)—female egg cell of oomycete fungi
oospore—thick-walled, sexually derived resting spore of oomycete fungi
ostiole (adj. ostiolate)—pore; opening by which spores are freed through the papilla or neck of a perithecium or pycnidium
ovary—ovule-bearing portion of a pistil
overwinter—to survive over the winter period
oviposition—egg-laying
ovule—immature seed
own-rooted vine—nongrafted vine, growing on its own adventitious roots

palisade—layer or layers of columnar cells rich in chloroplasts, beneath the upper epidermis of plant leaves
palmate—having lobes or veins radiating from a common point
papillum (pl. papilla, adj. papillate)—small, round or nipplelike projection
paragynous—having the antheridium at the side of the oogonium
paraphysis (pl. paraphyses)—sterile, upward-growing, basally attached hyphal element in a hymenium, especially in ascomycetes, in which they are generally clavate or filiform and branched or unbranched
parasite—organism that lives with, in, or on another organism (host) to its own advantage and to the disadvantage of the host
parenchyma—living cells, potentially capable of division, in an organ, such as a root, stem, leaf, or berry
parthenogenesis—development of an unfertilized egg into a new individual
pathogen (adj. pathogenic)—any disease-producing organism
pedicel—stalk of a flower or berry (in viticultural usage, called a cap stem)
peduncle—portion of a rachis (cluster stem) from the point of attachment to the shoot to the first lateral branch of the cluster
perfect—*see* hermaphroditic
perfect state—*see* teleomorph
pericyclic—pertaining to the outer part of the phloem of a stem or root, formerly thought to be distinct from phloem
periderm—*see* bark
peridium (adj. peridial)—outer envelope of the sporophore of many fungi
perithecium (pl. perithecia)—flask-shaped or subglobose, thin-walled ascocarp (fungal fruiting body), containing asci and ascospores and having an ostiole (pore) at the apex, through which spores are expelled or otherwise released
petal—one of the members of the cap of a grape flower
petiole—the stalk of a leaf
pH—negative logarithm of the effective hydrogen concentration, a measure of acidity (pH 7 is neutral; values less than pH 7, acidic; values greater than pH 7, alkaline)
phage—a virus parasitic on bacteria
phenological—of the relation of developmental stages of plants to seasonal changes
phialide (adj. phialidic)—end cell of a conidiophore; conidiophore of fixed length with one or more open ends, through which a basipetal succession of conidia develops
phloem—bast; food-conducting and food-storing tissue of roots, stems, etc.
photochemical oxidant—any of various highly reactive compounds formed by the action of sunlight on less toxic precursors
phyllody—change of floral organs to leaflike structures
physiologic race—subdivision within a species, the members of which are alike in morphology but differ from other races in virulence, symptom expression, biochemical and physiologic properties, or host range
phytoalexin—substance that inhibits the growth of certain microorganisms and that is produced in higher plants in response to a number of chemical, physical, and biological stimuli
phytotoxic—harmful to plants (usually used to describe chemicals)
phytotoxicity—injury or damage to a plant due to chemical treatment
phytotoxin—a toxin affecting plants
pistil—female structure of a flower, composed of the stigma, on which pollen grains germinate, and the style, through which the pollen tube grows to the ovule in the ovary
pith—loose, spongy tissue in the center of a stem (tissue composed of firm pith cells at a node is called a diaphragm)
plasmid—a self-replicating piece of DNA that is stably inherited in an extrachromosomal state
pollen—male sex cells produced by anthers of flowering plants
pollination—deposition of pollen on a stigma
pomace—skins, seeds, stems, and the like remaining after extraction of juice from grapes
primary bud—in viticultural usage, the central and largest bud in a compound winter bud, which usually develops into a fruiting shoot in the following growing season
primary infection—the first infection of a plant, usually in the spring by an overwintering pathogen
primary inoculum—inoculum, usually from an overwintering source, that initiates disease in the field, as opposed to inoculum that spreads disease during the season
primary symptom—the symptom produced soon after infection, in contrast to a secondary symptom, which follows more complete invasion
primordium (pl. primordia, adj. primordial)—the initial structure from which a plant part is formed
prokaryotic—lacking a nuclear membrane, mitotic apparatus, and mitochondria
propagule—any part of an organism capable of independent growth
prophyll—small, scalelike structure borne at the first one or more nodes of a stem
protectant—agent, usually a chemical, applied to a plant surface in advance of a pathogen to prevent infection
protoplasm—living contents of a cell
proximal—nearest to the point of attachment
pseudoaxillary—appearing to arise from the angle of a leaf with the stem, but not actually doing so
pseudoparenchymatous—fungal tissue consisting of closely packed, more or less isodiametric or oval cells resembling the parenchyma cells of higher plants
pseudothecium (pl. pseudothecia)—ascocarp similar to a perithecium and having a dispersed rather than an organized hymenium
pulvinate—cushionlike in form
pustule—blisterlike, small erumpent spot, spore mass, or sorus
pycnidiophore—specialized fungal hypha on which pycnidiospores are

produced

pycnidiospore—spore (conidium) produced in a pycnidium

pycnidium (pl. pycnidia)—asexual, globose or flask-shaped fruiting body of fungi producing conidia

pycnium (pl. pycnia)—the pycnidiumlike haploid fruiting body of rust fungi

pyriform—pear-shaped

quarantine—legislative control of the transport of plants or plant parts to prevent the spread of disease, pathogens, or other pests

rachis—main axis of a grape cluster

reservoir—an organism in which a parasite that is pathogenic to other species lives and multiplies without causing damage

resistance (adj. resistant)—property of hosts that prevents or impedes disease development

resting spore—temporarily dormant spore, usually thick-walled, capable of surviving adverse environments

rhizomorph—fungus mycelium arranged in strands, rootlike in appearance

rhizosphere—microenvironment in soil near and influenced by plant roots

rind—the firm outer layer of a rhizomorph or other organ

ring bark—bark shed as a sheath, or strips thereof

ring spot—disease symptom characterized by yellowish or necrotic rings enclosing green tissue, as in some plant diseases caused by viruses

rooting—a young vine produced from a rooted cutting grown for one season

rootlet—a slender root, especially a young lateral root

rootstock—rooted vine to which fruiting varieties of grapes are grafted to produce a commercially acceptable vine (grape rootstock varieties are used for their tolerance or resistance to root parasites, such as phylloxera and nematodes, or for their tolerance to soils)

rot—softening, discoloration, and often disintegration of succulent plant tissue as a result of fungal or bacterial infection

rugose—wrinkled

russet—yellowish brown or reddish brown scar tissue on the surface of fruit

safener—a chemical added to a pesticide to protect against phytotoxicity

sanitation—destruction of infected and infested plants or plant parts

saprophyte—nonpathogenic organism that obtains nourishment from the products of organic breakdown and decay

sapwood—outer part of xylem, in which active translocation of water and solutes takes place

scab—crustlike disease lesion

scion—fruiting variety that is grafted or budded onto a rootstock

sclerotium (pl. sclerotia)—hard, usually darkened and rounded mass of dormant hyphae with differentiated rind and medulla and thick, hard cell walls, which permit survival in adverse environments

scolecospore—fungal spore with a length-width ratio greater than 15:1

secondary bud—in viticultural usage, the second most developed of the three buds in a compound winter bud, adjacent to the leaf scar of the cane

secondary infection—infection resulting from the spread of infectious material produced after a primary infection or from other secondary infections without an intervening inactive period

secondary inoculum—inoculum produced by infections that took place during the same growing season

secondary organism—organism that multiplies in already diseased tissue but is not the primary pathogen

secondary rot—rot caused by a secondary organism

senesce (n. senescence, adj. senescent)—to decline with maturity or age, often hastened by stress from environment or disease

septum (pl. septa)—cross-wall

serology—a method using the specificity of the antigen-antibody reaction for the detection and identification of antigenic substances and the organisms that carry them

serration—sharp teeth pointing forward on a leaf margin

sessile—having no stem

shatter—*see* drop

shelling—abscission of flowers before or at bloom, occurring before drop

shoot—in viticultural usage, the succulent, green, current-season growth from a bud, including the leaves

shoot removal—deshooting; removal of unwanted shoots on the trunk of a vine below the head

shoot thinning—removal of unwanted shoot growth from the head, cordon, or arms of a vine when the shoots are short, usually 15–40 cm

shoot-tip culture—propagation of growing shoot tips, often done in vitro

shot berry—small, seedless grape berry that does not ripen

shot hole—disease symptom characterized by the dropping out of small, round fragments of leaves, making them look as if riddled by shot

shoulder—in viticultural usage, basal branch of a grape cluster

side-dressing—application of fertilizer or chemicals to the soil around established vines

sieve element, sieve tube—conducting cell of phloem

sign—indication of disease from direct visibility of a pathogen or its parts

sinus—cleft or indentation between the lobes of a leaf blade

slip-skin—skin of a ripe berry that separates readily from the flesh, characteristic of American grape cultivars such as Concord

softwood cutting—in grapevine propagation, a piece of shoot induced to produce adventitious roots

somatic—relating to the body, especially body cells as distinguished from reproductive cells

sorus (pl. sori)—compact fruiting structure of rust fungi

sp. (pl. spp.)—species (sp. used after a genus name refers to an undetermined species; spp. after a genus name refers to several species without naming them individually)

spermagonium (pl. spermagonia)—flask-shaped fungal structure producing sporelike bodies that may function as male gametes (spermatia); pycnium of a rust fungus

spermatium (pl. spermatia)—a sex cell; a nonmotile gamete

spermatization—the placing of spermatia on structures (receptive hyphae, etc.) for diploidization

sporangiophore—sporangium-bearing body of a fungus

sporangium (pl. sporangia)—fungal structure producing asexual spores, usually zoospores

spore—reproductive body of fungi and other lower plants, containing one or more cells; a bacterial cell modified to survive an adverse environment

sporulate—to produce spores

spur—in viticultural usage, a cane pruned to one to four nodes

stamen—male structure of a flower, composed of an anther, bearing pollen, and a filament, or stalk (*see also* pollen)

sterigma (pl. sterigmata)—small, usually pointed protruberance or projection

stigma—structure on which pollen grains germinate in a pistil

stipe—a stalk

stipule—small, short-lived structure formed at the base of the petiole of grape leaves

stock—*see* rootstock

stoma (pl. stomata, adj. stomatal)—structure composed of two guard cells and the opening between them, in the epidermis of a leaf, stem, or berry, functioning in gas exchange

stroma (pl. stromata)—compact mass of mycelium that supports fruiting bodies

stylar scar—small corky area remaining at the apex of a berry after the style dries and falls off following blossoming

style—*see* pistil

stylet—slender, tubular mouthparts in plant-parasitic nematodes or aphids

subcortical—beneath the cortex

subepidermal—beneath the epidermis

suberin—waxy, water-impervious substance deposited in or on plant cell walls

subglobose—not quite spherical

submicroscopic—too small to be seen with an ordinary light microscope

substrate—the substance on which an organism lives or from which it obtains nutrients; chemical substance acted upon, often by an enzyme

succulent—referring to a tender, juicy, or watery plant part

sucker—shoot arising from a trunk at or below ground level

summer lateral—*see* lateral shoot

symptom—indication of disease by reaction of the host

symptomless carrier—a plant infected with a pathogen (usually a virus) but having no obvious symptoms

syn.—synonym

synnema (pl. synnemata)—group of closely united and sometimes fused conidiophores, which bear conidia; syn. coremium

systemic—pertaining to a disease in which the pathogen (or a single infection) spreads generally throughout a plant; pertaining to chemicals that spread internally through a plant

teleomorph—the sexual form (also called the perfect state, or sexual stage) in the life cycle of a fungus, in which sexual spores (ascospores

or basidiospores) are formed after nuclear fission
teliospore—thick-walled resting spore produced by some fungi, notably rusts and smuts, that germinates to form a basidium
telium (pl. telia)—sorus that produces teliospores
tendril—twining organ borne opposite certain leaves of grapevine
tertiary bud—in viticultural usage, the least developed of the three buds in a compound winter bud, farthest from the leaf scar of the cane
tissue—group of cells, usually of similar structure, that perform the same or related functions
tolerance—capacity of a plant or crop to sustain disease or endure adverse environment without serious damage or injury
torulose—cylindrical but having swellings at intervals
toxin—poisonous substance of biological origin
trabecula (pl. trabeculae)—bar of wall-thickening material extending across the lumen from one tangential wall to the other of a conducting cell of xylem, as in certain virus-infected grapevines
translocation—movement of water, nutrients, chemicals, or elaborated food materials within a plant
transmit—to spread or transfer, as in spreading an infectious pathogen from plant to plant or from one plant generation to another
transpiration—water loss by evaporation from leaf surfaces and through stomata
trellis—permanent vine-supporting system consisting of posts, wire, and often crossarms
truncate—ending abruptly
trunk—in viticultural usage, the vertical aboveground axis of a grapevine
tuberosity—gall formed on a mature root by the feeding of phylloxera
tumorigenic—able to incite tumor formation
turgid—swollen, distended; referring to a cell that is firm due to water uptake
turgor pressure—internal pressure within plant cells, lack of which causes plants to wilt
tylosis (pl. tyloses)—evagination of a parenchyma cell membrane through a pit in an adjacent vessel (in grapevine, a tylosis eventually develops a thick wall; if numerous, tyloses may fill a vessel)

ultrastructural—relating to the structure of a cell as seen through an electron microscope
unitunicate—having one definable wall or cover
uredospore—repeating vegetative spore of rust fungi
uredium (pl. uredia)—fruiting body (sorus) of rust fungi that produces uredospores

vacuolate—referring to cavities in the cytoplasm containing fluids and dissolved substances
variegation—pattern of two or more colors in a plant part, as in a green and white leaf
variety—group of closely related plants of common origin and similar characteristics within a species (*see also* cultivar)
vascular—pertaining to conductive tissues (xylem and phloem)
vascular bundle—strand of conductive tissue, usually composed of xylem and phloem (in leaves, small bundles are called veins)
vector—agent that transmits inoculum and is capable of disseminating disease
vegetative—referring to somatic or asexual parts of a plant, which are not involved in sexual reproduction
vein—small vascular bundle in a leaf
veinbanding—discoloration or chlorosis occurring in bands along leaf veins, setting them off from interveinal tissue, a symptom of virus diseases
veinlet—small branch of a vein ending in the mesophyll of a leaf
véraison—loss of green color in grape berries beginning to ripen
vermiform—worm-shaped
verruculose—having small rounded processes or "warts"; delicately so
verticillate—whorled
vesicle (adj. vesiculate)—subcellular membranous enclosure
vessel—water-conducting cell of xylem
viable (n. viability)—able to germinate, as seeds, fungus spores, sclerotia, etc.; capable of growth
viroid—the smallest known infectious agent, consisting of nucleic acid and lacking the usual protein coat of viruses
virulent—pathogenic, capable of causing disease
viruliferous—virus-carrying (usually applied to insects or nematodes)

water-soaked—describing plants or lesions that appear wet and dark and are usually sunken and translucent
wilt—loss of freshness or drooping of plants due to inadequate water supply or excessive transpiration; a vascular disease interfering with water utilization
witches'-broom—disease symptom characterized by an abnormal, massed, brushlike development of many weak shoots arising at or close to the same point

xylem—water-conducting, food-storing, and supporting tissue of roots, stems, etc.

zonate—marked with zones; having concentric rings, like a target
zoospore—fungus spore with flagella, capable of locomotion in water

Selected References

Agrios, G. N. 1978. Plant Pathology, 2nd ed. Academic Press, New York. 703 pp.
Federation of British Plant Pathologists. 1973. A guide to the use of terms in plant pathology. Phytopathological Paper No. 17. Commonwealth Mycological Institute, Kew, Surrey, England.
Flaherty, D. L., Jensen, F. L., Kasimatis, A. N., Kido, H., and Moller, W. J., eds. 1981. Grape Pest Management. Publ. 4105. Division of Agricultural Sciences, University of California, Berkeley. 312 pp.
Hawksworth, D. L., Sutton, B. C., and Ainsworth, G. C. 1983. Ainsworth & Bisby's Dictionary of the Fungi. 7th ed. Commonwealth Mycological Institute, Kew, Surrey, England. 445 pp.
Office International de la Vigne et du Vin. 1963. Lexique de la Vigne et du Vin. OIV, Paris. 674 pp.
Raven, P. H., Evert, R. F., and Curtis, H. 1981. Biology of Plants, 3rd ed. Worth Publishers, New York. 686 pp.
Walkey, D. G. A. 1985. Applied Plant Virology. John Wiley & Sons, New York. 329 pp.

Index